Praise for
The Real Work of Data Science

These two authors are world-class experts on analytics, data management, and data quality; they've forgotten more about these topics than most of us will ever know. Their book is pragmatic, understandable, and focused on what really counts. If you want to do data science in any capacity, you need to read it.

Thomas H. Davenport
Distinguished Professor, Babson College and Fellow, MIT Initiative on the Digital Economy

I like your book. The Chapters address problems that have faced Statisticians for generations, updated to reflect today's issues, such as computational big data.

Sir David Cox
Warden of Nuffield College and Professor of Statistics, Oxford University

I am already in love with your book based on the overview and preface!! What a creative approach! Speaks a lot to your ability to tell a good story – one of the key ways of reasoning for a good data scientist!

Hollylynne S. Lee
Professor, Mathematics and Statistics Education and Faculty Fellow, Friday Institute for Educational Innovation, North Carolina State University

The root causes of business failures typically are management, not technology. In today's complex and changing digital world, the advice in *The Real Work of Data Science* is essential. Read it and do it.

John A. Zachman
Chairman – Zachman International and Executive Director – FEAC Institute

If you are wondering what the real challenges and solutions to solving your 'Big Data' problem are, this is a must read book. Ron and Tom move past the technology hype and highlight the real issues and opportunities in leveraging data science to the benefit of your organization

Jeff MacMillan
Chief Analytics and Data Officer, Morgan Stanley Wealth Management

Much needed!

Neil Lawrence
Professor of Machine Learning at the University of Sheffield and Machine Learning team manager at Amazon

More than 80% of data science projects fail, either partially or wholly, at the implementation stage. There is a wealth of books on the technical and mechanical aspects of data science, but little to guide data scientists and managers on the holistic integration of data science into organizations in a way that produces success. This well-written book fills that gap.

Peter Bruce
Founder and Chief Academic Officer, The Institute for Statistics Education

C'est livre est très intéressant et plein de très bonnes choses intelligentes et utiles. Il sera sans nul doute très précieux.

Jean Michel Poggi
Professor of Statistics at Paris-Descartes University and Mathematics Laboratory, Orsay University, Paris, France, Past President of the Société Française de Statistique and Vice-President of the Federation of European National Statistical Societies

I like the very direct and succinct style. You are certainly right on target when you say you can't stress enough the importance of understanding the real problem. Other of your points in Chapter 1 really hit home, such as data scientists spending more time on data quality than on analysis. (I'm glad they do.) Further, you are absolutely correct that data scientists must translate their results into the language of the decision-maker. I also recognize the liberal use of anecdotes in the book. For instance, the remarks about Bill Hunter, the ice cream sales, the Pokémon experiment, etc. I personally like this, and I do this in all of my speeches since I think it really hooks the audience.

Barry Nussbaum
Past Chief Statistician, the United States Environmental Protection Agency and Past President of the American Statistical Association

I think this book is excellent for an introductory course in data science. It could be used with students at university level or with professionals in specialist courses.

Luciana Dalla Valle
Lecturer in Statistics and Programme Manager of the MSc Data Science and Business Analytics, School of Computing, Electronics and Mathematics, Plymouth University, UK

The Real Work of Data Science addresses the softer issues of data science that actually decide on the success or failure of any data science initiative. It makes the data science and Chief Analytics Officer roles more understandable and accessible to a wider audience. Choosing the right modeling method is often the key point of discussion in books, although it is just a tiny fraction of the job to be done. This book prepares you for the harsh reality of data science in the real-world.

Alexander Borek
Global Head of Data & Analytics at Volkswagen Financial Services

Data science is critical for competitiveness, for good government, for correct decisions. But what is data science? Kenett and Redman give, by far, the best introduction to the subject I have seen anywhere. They address the critical questions of formulating the right problem, collecting the right data, doing the right analyses, making the right decisions, and measuring the actual impact of the decisions. This book should become required reading in statistics and computer science departments, business schools, analytics institutes and, most importantly, by all business managers.

A. Blanton Godfrey, Joseph D. Moore
Distinguished University Professor, Wilson College of Textiles, North Carolina State University

THE REAL WORK OF
DATA SCIENCE

THE REAL WORK OF DATA SCIENCE

TURNING DATA INTO INFORMATION, BETTER DECISIONS, AND STRONGER ORGANIZATIONS

Ron S. Kenett

Ra'anana, Israel

Thomas C. Redman

Rumson, NJ, USA

This edition first published 2019
© 2019 Ron S. Kenett and Thomas C. Redman

All rights reserved. No part of this publication may be reproduced, stored in a retrieval system, or transmitted, in any form or by any means, electronic, mechanical, photocopying, recording or otherwise, except as permitted by law. Advice on how to obtain permission to reuse material from this title is available at http://www.wiley.com/go/permissions.

The right of Ron S. Kenett and Thomas C. Redman to be identified as the authors of this work has been asserted in accordance with law.

Registered Offices
John Wiley & Sons, Inc., 111 River Street, Hoboken, NJ 07030, USA
John Wiley & Sons Ltd, The Atrium, Southern Gate, Chichester, West Sussex, PO19 8SQ, UK

Editorial Office
9600 Garsington Road, Oxford, OX4 2DQ, UK

For details of our global editorial offices, customer services, and more information about Wiley products visit us at www.wiley.com.

Wiley also publishes its books in a variety of electronic formats and by print-on-demand. Some content that appears in standard print versions of this book may not be available in other formats.

Limit of Liability/Disclaimer of Warranty
While the publisher and authors have used their best efforts in preparing this work, they make no representations or warranties with respect to the accuracy or completeness of the contents of this work and specifically disclaim all warranties, including without limitation any implied warranties of merchantability or fitness for a particular purpose. No warranty may be created or extended by sales representatives, written sales materials or promotional statements for this work. The fact that an organization, website, or product is referred to in this work as a citation and/or potential source of further information does not mean that the publisher and authors endorse the information or services the organization, website, or product may provide or recommendations it may make. This work is sold with the understanding that the publisher is not engaged in rendering professional services. The advice and strategies contained herein may not be suitable for your situation. You should consult with a specialist where appropriate. Further, readers should be aware that websites listed in this work may have changed or disappeared between when this work was written and when it is read. Neither the publisher nor authors shall be liable for any loss of profit or any other commercial damages, including but not limited to special, incidental, consequential, or other damages.

Library of Congress Cataloging-in-Publication data has been applied for

ISBN: 9781119570707

Cover Design: Wiley
Cover Image: © enisaksoy/Getty Images

Set in 10/12pt Times by SPi Global, Pondicherry, India

Printed and bound by Quad/Graphics in the United States of America

V252436_051519

To Sima, our children and their families, and their wonderful children: Yonatan, Alma, Tomer, Yadin, Aviv, Gili, Matan, Eden and Ethan

Ron

To my wife Nancy, our six children, and our grandchildren

Tom

Contents

About the Authors

Prof. Ron S. Kenett is Chairman of the KPA Group and Senior Research Fellow at the Samuel Neaman Institute, Technion, Haifa, Israel. He is an applied statistician combining expertise in academic, consulting, and business domains. Ron is past president of the Israel Statistical Association and the European Network for Business and Industrial Statistics. He has written more than 250 papers and 14 books on statistical methods and applications. He was awarded the 2013 Greenfield Medal by the English Royal Statistical Society and the 2018 Box Medal by the European Network for Business and Industrial Statistics in recognition of excellence in contributions to the development and application of statistics.

Dr. Thomas C. Redman, "the Data Doc," President of Data Quality Solutions, helps start-ups, multinationals, senior executives, chief data officers, and leaders buried deep in their organizations chart their courses to data-driven futures, with special emphasis on quality and data science. The author of five other books and hundreds of papers, Tom's most important article is "Data's Credibility Problem" (*Harvard Business Review*, December 2013). He has a PhD in statistics and two patents. Tom lives in Rumson, New Jersey, with his wife, Nancy.

Preface

This book has its roots in a chance meeting brought on when Ron responded to an article on data science that Tom published. One short discussion led to another, quickly narrowing to a common theme: we shared the experience that, in order to help companies and organizations become better at exploiting data and statistical analysis, one needs something more than technical brilliance. For both of us, our most successful and impactful projects resulted from other factors, such as understanding the problem, narrowing the focus, delivering simple messages in powerful ways, being in the right spot at the right time, and building the trust of decision-makers. Conversely, our failures stemmed not from poor technical work but from a failure to connect, on the right issues, with the right people, or in the right way.

We had both written, separately, on some aspects of these topics. Ron has studied how one generates information quality with a framework labeled "InfoQ," Tom has addressed data quality and became known as "the Data Doc." We wondered if we could help data scientists who work in companies and other organizations enjoy more and larger successes and endure fewer failures by putting our heads together.

Fad, Trend, or Fundamental Transformation?

It is no secret that "data," broadly defined, is all the rage. And "data science," including traditional statistics, Bayesian statistics, business intelligence, predictive analytics, big data, machine learning, and artificial intelligence (AI) are enjoying the spotlight. There are plenty of great successes, building on a rich tradition of statistics in government and industry, driven by increasing business needs, more data powered by social media, the Internet of Things, and the computer power to analyze it. Iconic new companies include Amazon, Facebook, Google, and Uber. At the same time, there are enormous issues: the Facebook/Cambridge Analytica scandals of early 2018 underscore threats to our privacy (Kenett et al. 2018), many fear that millions of jobs will be lost to artificial intelligence, analytics projects still fail at a high rate, and the tremendous damage that has resulted from some notable "successful" efforts, as described in O'Neil (2016).

Will data and data science power the next great economic miracle? Will they make solid contributions, more positive than negative? Or will they be just another fad confined to the scrap heap of failed ideas? Even worse, will they put our entire social fabric at risk? It is impossible to know.

We do know that data and data science can be truly transformative, improving customer satisfaction, increasing profits, and empowering people – we have seen it with our own eyes.

We believe that data scientists have huge roles to play in tipping the scales toward the good in the questions above. This will require incredible commitment, determination, and follow-through. We encourage data scientists, statisticians, and those who manage them to take up the cause, as we have. We want to do all we can to fully equip them.

Data Scientists and Chief Analytics Officers

In writing the book, we adopted four "personas" as readers. First is Sally, a 31-year-old data scientist who works in a midsize department or company. Sally's job involves producing management reports, although she does have some time for teasing insights from ever-increasing volumes of untrustworthy data. Her title could be any of "data scientist," "statistician," "analyst," "machine learning specialist," and others. We are well aware that some people see differences between these titles. But (with one exception, below) those distinctions are meaningless for us. Whether you are trained as a statistician, computer scientist, physicist, or engineer, your job is to turn "data into information and better decisions," as part of our title demands.

Our second reader persona is Divesh, the 50-year-old who has the top analytics job within his department, business unit, or company. His title may be "chief analytics officer," "head of data science," or something similar. Divesh may have no formal training in data science, but he is a seasoned manager. While Divesh's day job is to manage data science across his department, within his sphere, he also bears special responsibility for the "building stronger organizations" portion of our title.

Brian, a solid industrial statistician, aged 46 and employed as an internal consultant, is our third persona. Brian is simultaneously bemused and threatened by data science, and he sits on the sidelines way too much. We think Brian has much to offer and encourage him to join the effort.

A fourth persona has an outsized impact on data science and this book. It is Elizabeth, who heads up some department, division, even an entire company. Liz hated statistics in college – it was a required course, poorly taught, and not connected to the rest of her studies. She has seen more and more power in data and data science over the last several years and is just beginning to explore what it means for her department. Liz is both excited about the possibilities and fearful that her efforts will fail miserably.

More than anything, Liz's success, or failure, will dictate the future of data science. She can ignore it (and there are plenty of good reasons to do so) or become an increasingly demanding customer. If she fully embraces data and data science, she can transform her department.

Introduction to the Book

Sally, Divesh, and Brian have different needs but share a common theme. Their business is to turn numbers into information and insights. To be useful, their analyses need to guide decisions that carry a positive impact in the workplace. In other words, they need to help Liz succeed.

We packaged our experience in 18 short chapters directly relevant to our four main personas. We do not deal with technical issues but instead focus on the *make or break* ingredients in data-driven transformation.

The chapters cover the different steps data scientists take in organizations. We discuss their role as individuals and through their organizational positions. We present lots of models that have helped us, we discuss the integration of hard and soft data in analytic work, and we stress the importance of impact (as opposed to technical excellence). The book also provides a context and opens curtains to landscapes that are not usually explored by most experts in data analysis.

We build on the contributions of statisticians like Box, Breiman, Cox, Deming, Hahn, and Tukey; cognitive psychologists like Kahneman and Tversky; and leaders in other disciplines to address current and future challenges. We also connect theory and applications, past contributions and modern developments, organizational needs and the means to fulfill them.

We've been as direct and to the point as we are able. This book should help you think more broadly about your job. Those seeking cookbook style "how-tos" will be sadly disappointed. It does provide an overview, benchmarks, and objectives, but you will have to develop your own concrete action plans.

We will be successful if readers take ideas introduced here and apply them in ways that best suit their own skill sets, the needs of decision-makers they serve, and the cultures of their organizations. Data and analytics can transform organizations for the good – we encourage data scientists and applied statisticians to do their part, to help decision-makers become more effective, and to keep this transformation on the right track.

About the Companion Website

This book is accompanied by a companion website:

www.wiley.com/go/kenett-redman/datascience

The website material includes:

• A List of Useful Links

Scan this QR code to visit the companion website.

1

A Higher Calling

It is a great time for data science! The *Economist* proudly proclaims that data is "the world's most valuable resource,"[1] and Hal Varian and Tom Davenport[2] have variously called statistics and data science "the sexiest job of the twentieth century." In searching the web for the term *data scientist*, we find the following definition, *"'Data Scientist' means a professional who uses scientific methods to liberate and create meaning from raw data."*[3] Similar definitions have been offered for statisticians and data analysts.[4] Yet we believe the work is more involved and requires skills far beyond those needed to create meaning from raw data.

This book expands and clarifies what it takes to succeed in this job, within the organizational ecosystem in which it takes place. It builds on years of experience in a wide range of organizations, all over the world. Our goal is to share this experience and some retrospective insights learned in doing real work. Specifically, we propose that the real work of data scientists and statisticians involves helping people make better decisions on the important issues in the near term and building stronger organizations and capabilities in the long term. By "people" we mean, among others, managers in organizations and professionals in service and production industries. This perspective is also relevant to educators in schools and colleges and researchers in laboratories and academic institutions. It is a far higher, and more demanding, calling. For example, you don't get to contribute on the really important decisions unless you're trusted.

Thus, the real work requires total involvement: helping to formulate the problems and opportunities in crisp business or scientific terms; understanding which data to consider and the strengths and limitations in the data; determining when new data is needed; dealing with quality issues; using the data to reduce uncertainty; making clear where the data ends and intuition must take over; presenting results in simple, powerful ways; recognizing that all important decisions involve political realities; working with others; and supporting decisions in practice. This real work is not taught enough in statistics or data science courses.

---

[1] Cover of the May 6, 2017, issue.
[2] https://hbr.org/2012/10/data-scientist-the-sexiest-job-of-the-21st-century (Davenport and Patil 2012).
[3] http://www.datascienceassn.org/code-of-conduct.html (Data Science Association 2018).
[4] Herein, we use the terms *data science*, *data analytics*, and *statistics* interchangeably, fully recognizing that many people see fine distinctions. But these are not central to this book.

The Real Work of Data Science: Turning Data into Information, Better Decisions, and Stronger Organizations, First Edition. Ron S. Kenett and Thomas C. Redman.
© 2019 Ron S. Kenett and Thomas C. Redman. Published 2019 by John Wiley & Sons Ltd.
Companion website: www.wiley.com/go/kenett-redman/datascience

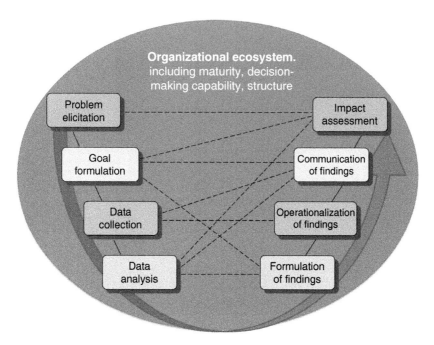

Figure 1.1 The life-cycle view of data analytics, in the context of the organizational ecosystem in which the work takes place.

The unpleasant reality is that many/most companies derive only a fraction of the value that their data, data science, and statistics offer (see, for example, Henke et al. 2016). Data scientists and their managers, including chief analytics officers (CAOs), chief data scientists, heads of data science, and other professionals who employ data scientists,[5] must learn how to address the barriers that get in the way. Thus, the real work also involves raising everyone's ability to conduct simple analyses and understand more complex ones, understand the power of data, understand variation, and integrate data and their intuitions; putting the right data scientists and statisticians in the right spots; educating senior leadership on the power of data; helping them become good consumers of data science; teaching them their roles in advancing the effort; and creating the organizational structures needed to do all of the above effectively and (reasonably) efficiently. This is what this book is about.

Providing the added value we are talking about requires a wide perspective. Figure 1.1 presents the life cycle of data analytics in the context of an organization aiming to profit from data science (adapted from Kenett 2015). As the figure illustrates, the work is highly iterative (for more on this process, see Box 1997).

The Life-Cycle View

The life-cycle view is designed to help data scientists help decision-makers. Let's consider each step of the cycle in turn.

---

[5] We recognize once again that many people see distinctions in these roles as well, but we will also use them interchangeably, as the distinctions are not central to this book.

Problem Elicitation: Understand the Problem
Observe what happens when you go to a dentist: you give a dentist a hint about your symptoms, you are placed in the chair, the dentist looks into your mouth, diagnoses and (hopefully) solves the problem, and tells you when to come back, all in less than an hour.

The seasoned data scientist knows better. We describe these data scientists in Chapter 2. They listen carefully and ask probing questions, keeping the customers (e.g. the decision-makers) focused and obtaining the relevant details to understand their needs. It may be an operations manager experiencing huge costs because of rework, a marketing manager trying to enter a new market, or a human resources (HR) manager who wants to reduce employee turnover. The experienced data scientist also reads the customer's body language for unspoken clues: does the customer have a hidden agenda, is he or she trying to make someone else look bad or build support for a political squabble?

Like many others, we can't stress this enough – you simply must understand the real problem if you hope to help solve it. The quality of analytic work depends on it (Kenett and Shmueli 2016a). More in Chapters 3 and 4.

Goal Formulation: Clarify the Short-term and Long-term Goals
Don't expect that the decision-maker has clearly formulated the problem. Bill Hunter, a famous statistician from the University of Wisconsin in Madison, tells the story of two chemists who sought his advice. When he asked them to describe their problem, they entered a lengthy discussion that led them to reformulate their problem. This one was much simpler, and they did not need further help from Bill. They left his office after thanking him profusely (Hunter 1979). While Bill's role may seem small, it was essential!

The main point is that a full understanding of the problem requires a full understanding of the context in which it occurs, including the overarching goal. More in Chapter 4.

Data Collection: Identify Relevant Data Sources and Collect the Data
Cobb and Moore (1997) point out that "Statistics requires a different kind of thinking, because data are not just numbers, they are numbers with a context." The context helps identify relevant data sources and their interpretation.

To illustrate, consider this story from Denmark from Kenett and Thyregod (2006). It involves an exercise in a fourth-grade textbook and shows the importance of context and how numbers turn into data. In this exercise, the numbers presented in Figure 1.2 record the number of ice creams sold each day, without any indication of the actual day of the week. In July, it was very hot for nine consecutive days. Students were asked to (i) identify the hot days and (ii) determine which days were Sundays.

By itself, the graph just presents 31 numbers. But Danish schoolchildren know their parents are more inclined to offer ice cream on weekends and on hot days. With this context, it was easy for these young children to complete their assigned tasks.

Context is revealed where data is generated, from the shop floor, to the laboratory, to a social media setting. Data scientists must understand this context and identify the data relevant to the problem. More on this in Chapter 5.

Data Analysis: Use Descriptive, Explanatory, and Predictive Methods
This is the work of "creating meaning from data," "separating the signal from the noise," "turning data into information," and so forth. There are, of course, literally thousands of

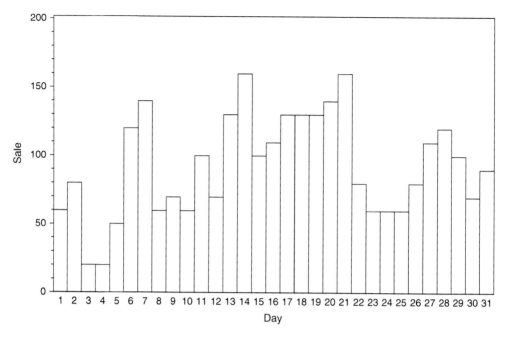

Figure 1.2 The number of ice creams sold in a Danish locality, by day in July.

examples. As one, consider eBay auctions. When you sell an item on eBay, you are asked to specify a "reserve price," a value you set to start the auction. If the final price does not exceed the reserve price, the auction does not transact. On eBay, sellers can choose to place a *public* reserve price that is visible to bidders or a *secret* reserve price (bidders only see that there is a reserve price but do not know its value).

Katkar and Reiley (2006) investigated the effect of this choice. Their data came from an experiment selling 25 identical pairs of Pokémon cards, where each card was auctioned twice, once with a public reserve price and once with a secret reserve price, and consists of complete information on all 50 auctions. They used linear regression and significance tests to quantify the effect, if any, of private/public reserve on the final price. They concluded that "a secret-reserve auction generates a $0.63 lower price on average," a simple statement everyone can understand.

We are less concerned with this work here, except for one critical area usually not well covered in data science training. The cold, brutal reality is that too much data is unfit for analysis (Nagle et al. 2017), and data scientists spend far more of their time on data quality issues than they do on analysis. High-quality data is critical for all analyses and especially so for cognitive technologies (Redman 2018b). So data scientists must deal with the issue. More in Chapter 6.

Formulation of Findings: State Results and Recommendations
Analytics produces outputs such as descriptive statistics, p-values, regression models, analysis of variance (ANOVA) tables, control charts, trees, forests, neural networks, dendrograms, and

so forth. Many are beyond the scope for decision-makers. So, it is essential that data scientists translate their results into language the decision-maker understands.

Further, data scientists must explore the implications of their results and, oftentimes, recommend specific courses of action. Said differently, data scientists cannot simply "throw results and recommendations over the wall." Rather, they must ensure that the decision-maker understands the findings in their proper context. Because many people are involved in important decisions, this may mean several distinct presentations and interactions with senior managers, middle managers, and knowledge workers. All may require different levels of detail, in different forms.

Concepts and notation from mathematical statistics turn many people off. Instead, well-thought-out graphical displays are the communication tools of choice. Findings that cannot be presented in a graph are probably not worth communicating. Keep graphs and slides simple, and keep the "ink-to-information ratio" low, avoiding fancy symbols etc. (Tufte 1997). For a simple example, see Chapter 7.

A great example of this involves an analyst who realized that senior decision-makers did not understand the technical terms associated with the network robustness problem they assigned him. So, he crystallized the problem and formulated his results using a well-known fairy tale: "The first thing we must decide is what kind of network we want: a 'baby bear network,' a 'mama bear network,' or a 'papa bear' network. Roughly this means ..." Everyone got it.

While the actual decision is made by others, in the life-cycle model we expect the analysis to support a decision, even a tentative one, as the conclusion to this step.

Operationalization of Findings: Suggest Who, What, When, and How
The data scientist's job does not end with a decision. Rather, he or she should follow the data-based decisions into execution, helping define how results are put into practice (e.g. operational procedures), answering questions that are sure to come up, evaluating new data as it comes in, and advising on situations beyond the scope of the original analysis.

It is tempting to skip this step. But the value of data science only accrues when an analysis and decision are put into practice, not before. More in Chapter 8.

Communication of Findings: Communicate Findings, Decisions, and Their Implications to Stakeholders
Until now, a relatively small number of people have been involved in the work we referred to. But important decisions can impact thousands, even millions, of people. At this step, findings must be communicated to all who may be impacted, a much wider audience than those involved in the decision. While the lion's share of this work is the purview of the decision-maker, the data scientist must play an active role in support.

Impact Assessment: Plan and Deploy an Assessment Strategy
Although it is beyond the scope of helping decision-makers per se, data scientists should assess their impact. Wherever possible, get hard numbers. Of course, as the vignette featuring Bill Hunter illustrates, this is not always possible. And even when you can get hard numbers, solicit feedback from decision-makers.

Then be brutally honest in assessing how you can do better next time. More in Chapter 9.

The Organizational Ecosystem

The work of data science takes place in complex organizational settings, which can both promote and limit its effectiveness. Sometimes simultaneously. Data scientists and CAOs must be aware of and, over time, improve several components of the overall "organizational ecosystem."

The term *data-driven* has invaded the lexicon. One sees extravagant claims for data-driven marketing, data-driven HR, and data-driven technologies. Beyond the hype, and more deeply, is a powerful core that leads to better decisions and stronger organizations. At that core, the more data-driven the organization, the more demanding decision-makers are of data scientists, the more seriously they take sophisticated analyses, and the more they invest in high-quality data, clear decision rights, and the decision-making capabilities. Thus, smart data scientists and CAOs invest considerable time in educating themselves and decision-makers at all levels about this powerful concept and working together to advance it across the organizations.

We will discuss what it means to be data-driven in some detail in Chapter 10. It will come as no surprise that bias, in any form, is diametrically opposed to data-driven decision-making. Step one for data scientists is to remove bias from their own work – a subject we will take up in Chapter 11. The focus of Chapters 12–14 is education. First, Chapter 12 advises data scientists to start with the basics with their peers and other decision-makers. Chapter 13 takes a slightly different tack. It recognizes that demanding customers (e.g. decision-makers) will do as much to advance a data-driven culture and data science as anything. So, the chapter provides a list of questions to help decision-makers know what to ask.

With big data, AI, security concerns, the General Data Protection Regulation (GDPR), digitization, and so much more all over the news, it is hard for senior leaders to see the data space in perspective. Chapter 14 considers the big picture, advising CAOs to develop a wide and deep perspective on the data space and to help their organization's most senior leaders understand the risks and opportunities.

Organizational Structure
The unfortunate truth is that where data scientists sit in an organization dictates what they can do. For example, a data scientist sitting in the maintenance department may be denied access to relevant data from the operations department, for no other reason than the heads of each are competing for the same promotion. While data scientists may like to believe they are above it all, there is no escaping politics. Better for data scientists and CAOs to embrace this reality and strive to get into the right spots. More on this in Chapter 15.

Organizational Maturity
Finally, organizations have different needs of data science, based on their maturity. These run the gamut from those in fire-fighting mode, with basic, immediate needs, to learning organizations with needs for deep, penetrating analyses and predictions. More in Chapters 16 and 17.

Once Again, Our Goal

With this background, our goal is to help data scientists and CAOs become more effective. This means helping data scientists contribute to better decisions and CAOs to stronger organizations, without being too strict about it. We have organized the material as 18 narrowly focused chapters, loosely tied to the life cycle of analytics and the ecosystem. We have also included some materials not directly tied to either. For example, the next chapter explores the difference between a good data scientist and a great one. Each chapter is short and to the point.

Overall, this book presents a wide-angle view of the real work of data science. Our goal is to expand your perspective, trigger deep thinking, and help you develop insights that make you more effective as a developer, participant, or consumer of the data science experience.

2

The Difference Between a Good Data Scientist and a Great One

The difference between a good data scientist and a great one is like the difference between a lightning bug and lightning.* Indeed, they are two separate beasts.

Good data scientists work to discover hidden insights in vast quantities of often disparate and often poor-quality data. It is a demanding job. Still, good data scientists discover new insights into customer needs, the causes of variability in processes, and how the business is performing that others cannot. They are rare and extremely valuable contributors.

Great data scientists think about things differently. They are not simply interested in finding new insights in the data. They are interested in developing new insights about the larger world around them. Of course, they use the data to do so. But they also use anything else they can get their hands on.

To illustrate, consider predictions for the winner of the 2016 presidential election. As of November 7, 2016, pollsters predicted a Clinton victory over Trump with high probability:

Pollster	Probability of Clinton win
538 (Nate Silver):	72%
New York Times:	86%
Princeton Election Commission (PEC):	>99%

Importantly, none of these pollsters actually conducted polls themselves. Instead they built models using the raw data provided by others. We're impressed with Nate Silver and 538 and, to his credit, Mr. Silver acknowledged the weaknesses in polling in his final note just before the election. Although we do not know who did the work, we are confident that the *Times* and the PEC employed good data scientists. For a review of election surveys, see Kenett et al. (2018).

---

*This Chapter is based, in part, on a pair of *Harvard Business Review* digital articles by Redman (2013a, 2017a).

The Real Work of Data Science: Turning Data into Information, Better Decisions, and Stronger Organizations, First Edition. Ron S. Kenett and Thomas C. Redman.
© 2019 Ron S. Kenett and Thomas C. Redman. Published 2019 by John Wiley & Sons Ltd.
Companion website: www.wiley.com/go/kenett-redman/datascience

Great data scientists cast a much deeper and wider net. They "go deep" by studying past polls to get a get a sense of their strengths and weaknesses. In doing so, they will have learned (for example) that people lie to pollsters. In mixed company, not a single person we hang out with confessed that he or she planned to vote for Trump. But privately many admitted, "I'm going to vote for Trump. I just don't want my wife [or husband] to know."

Similarly, a few in the media commented on how much more energy they felt at Trump rallies than at Clinton rallies. They concluded that those who said they were going to vote were more likely to actually do so. Even a small amount of lying or misplaced optimism about voting could skew poll results. The great data scientist will conduct some simple simulations to learn more.

Further, there are plenty of other predictors of presidential victors, based on the economy, the rate of employment, the winner of the previous Super Bowl, and so forth. Thus a great data scientist will "go wide" also. To illustrate, some note that Americans eschew political dynasty. So, after one party has held the presidency for two terms, Americans will lean toward the other. Prior to 2016, we count eight relevant elections, the "other party" having won six. By this logic, one would estimate the probability of a Trump victory as $6/8 = 75\%$.

Note that great data scientists are not simply searching for the single best set of data, explanation, or model. They are seeking to understand many perspectives, to see which support one another, which conflict, how much variation they portend, and anything else that bears. They talk to all sorts of people, try out new theories, ruthlessly discard those that do not satisfy, and are always on the lookout for more and different data. This is how they find out the way the world works!

Appendix A lists some of the traits of such data scientists.

Over the years we've had the privilege of working with dozens, maybe hundreds, of good data scientists, statisticians, and analysts. And a few great ones. This relentless focus on learning about the world is the key differentiator. The great ones possess four other traits as well:

1. *They grow and take advantage of large networks*. They need them. They are interested in many things and can't possibly be expert in all of them. Great data scientists cultivate relationships with people who have different perspectives than their own. So much the better to explore the world, learn of new sources of data, and try out interim theories.

2. *They have a certain quantitative knack*. Great data scientists simply see things that others don't. For example, a summer intern (who now uses his analytical prowess as head of a media company) on his second day at an investment bank exhibited this inherent capability. His boss had given him a stack of things to read, and in scanning through, he spotted an error in a return's calculation. It took him about an hour to verify the error and determine the correction.

   What's important here is that thousands of others did not see the error. It was obvious to him, but not to anyone else. And this was a top-tier investment bank. Presumably, at least a few good analysts read the same material and did not spot it. Mathematics has turned out to provide a convenient, amazingly effective language (Einstein used the phrase "unreasonably effective") for describing the real world. The great data scientist taps into that language intuitively and easily in ways that even good data scientists cannot.

3. *They have persistence*. The great data scientists are persistent, and in many ways. The intern in the vignette above made his discovery at a glance and confirmed it in an hour.

It rarely works out that way. As Jeff Hooper, of Bell Labs, liked to say, "Data do not give up their secrets easily. They must be tortured to confess."

This is a really big deal. Even under the best of circumstances, too much data is poorly defined and simply wrong, and most turns out to be irrelevant to the problem at hand. Staring through this noisy data is arduous, frustrating work. Even good data scientists may move on to the next problem. Great data scientists stick with it.

Great data scientists also persist in making themselves heard. Dealing with a recalcitrant bureaucracy can be even more frustrating than dealing with noisy data. Continuing the vignette from above, the intern spent his summer defending his discovery. Whichever group made the error took great offense, even attacking him personally. Others reacted with glee as they celebrated the ignorance of their peers. And he was caught in the middle. Great data scientists know how to handle such situations, persisting through thick and thin.

4. *Finally, they have raw statistical muscle.* The abilities to access and analyze data using all the newest tools (including classic packages and newer ones such as machine learning) are obviously important. But these can learned – of bigger concern is the ability to bring statistical rigor to bear. At the risk of oversimplifying, there are two kinds of analyses – descriptive and predictive. Descriptive analyses are tough enough. But the really profitable analyses involve prediction, which is inherently uncertain (Shmueli 2010).

Great data scientists embrace uncertainty. They recognize when a prediction rests on solid foundations and when it is merely wishful thinking. They are simply outstanding in describing what has to go right for the prediction to hold, what will really foul it up, and what are the unknowns that keep them awake at night. When they can, they quantify the uncertainty, and they are good at suggesting simple experiments to confirm or deny assumptions, reduce uncertainty, explore the next set of questions, etc.

To say this in a different way, there are some who opine that, for big data, it is enough to understand "correlation" without getting into the complexities of "causation." There are surely some problems for which this is true. But not the really important ones! Understanding causation leads to better predictions. The great data scientists will work to establish the causative links.

This requires them to generalize on a higher level. Focusing only on the data at hand can lead to "overfitting," leading to models that are too complex for future use. Scientific generalization invokes domain-specific knowledge, general principles, and intuition, far beyond cross-validations or comparison of training-set and hold-out-set results (Kenett and Shmueli 2016a).

To be clear, this ability is not "that certain quantitative knack." It is trained, sophisticated, disciplined inferential horsepower, practiced and honed by both success and failure.

Some of this is covered in data science curricula (De Veaux et al. 2017; Coleman and Kenett 2017). Most is not.

Implications

To conclude, the real work of a data scientist is to continually become more effective. You probably cannot teach yourself "that certain quantitative knack." But you can work to develop outside interests, read extensively, build a wider, more diverse network, develop a thick skin, and study statistical inference. You should start doing so immediately.

The real work for chief analytics officers is a bit more involved, and we will explore it in later chapters. For now, understand that great data scientists are truly special. They are the Derek Jeters, the Michael Jordans, the Mikhail Barishnikovs, and the Julia Robertses of the data space. If you're serious about AI, big data, and advanced analytics, you need to find one or two, build around them, and craft an environment that helps them do their things.

3

Learn the Business

To be effective, data scientists should learn all they can about the company or agency that employs them, including its customers, products, and services; the overall industry, including major partners and competitors; and management's vision for the future. Some companies provide good orientation programs for new employees. Even so, learning the business is a lifelong proposition. There are many ways to learn about a business – here are some options.

The Annual Report

A good place to start is the company's annual report. The document is usually prepared with shareholders and capital market analysts in mind, yet it can be also informative for data scientists. The management introduction describes the company's vision, its short- and long-term goals, what it accomplished in the past year, and where it is headed in the next few.

Annual reports can produce some important surprises. As an example, can you guess Steve Job's original vision for Apple?

Answer: While reading a *Scientific American* article on the relative efficiency of animals, Steve Jobs was struck by how ordinary human beings are. For example, we are in the middle of the pack in converting energy to speed. But put a human on a bicycle, and the human far exceeds the efficiency of any animal. Steve's vision was for Apple to become a *"bicycle for the mind."* This vision drove its product and service line roadmap.

While many missions and visions are quite pedestrian, for some organizations they serve as a unifying principle and raison d'être and drive various strategic initiatives.

SWOTs and Strategic Analysis

A classical approach to evaluate a business or any organization is to list its strengths, weaknesses, opportunities, and threats, i.e. SWOT. In such a mapping, strengths and weaknesses represent an inward look and opportunities and threats represent the outside landscape. An infinite number of management meetings have been dedicated to SWOT analyses. Studying an organizational SWOT is an excellent way to learn about a company (Kenett and Baker 2010).

The Real Work of Data Science: Turning Data into Information, Better Decisions, and Stronger Organizations,
First Edition. Ron S. Kenett and Thomas C. Redman.
© 2019 Ron S. Kenett and Thomas C. Redman. Published 2019 by John Wiley & Sons Ltd.
Companion website: www.wiley.com/go/kenett-redman/datascience

Strategic analysis, beyond SWOTs, usually focuses on potential new projects or business propositions. This can cover a wide scope, such as:

- fit with business or corporate strategy
- inventive merit
- strategic importance to the business
- durability of the competitive advantage
- reward based on financial expectations
- competitive impact of technologies
- probability of success
- R&D cost to completion
- time to completion
- capital on hand and marketing investment required to exploit new opportunities
- effect on market segments
- implications to product categories or product lines.

Sort Out the Value Structure

An organization's values play key roles in shaping its direction, choice of metrics, and the decisions made by individuals and groups. Redman learned this in one of his first consulting assignments, with a large investment bank. His assignment involved helping the company sort out its data quality program, and he pitched the program as saving money. But he got little traction.

A chance event involving a Super Bowl pool helped Redman see that, even though the bank carefully tracked expenses, saving money was not high on its list of priorities. Rather, the bank prided itself on growing revenue and managing risk. Recasting the data quality program along these lines helped move it along.

The vignette illustrates a more general point: it takes more to understand a company than studying its formal documents. Of particular concern to data scientists is who makes the important decisions (e.g. senior or more junior people), how they are made (e.g. by consensus or by the most senior person), and under what criteria (e.g. driving revenue, increasing shareholder value, improving customer satisfaction, regulatory concerns, innovation, etc.).

Data scientists should ask to see these studies (even participate in their creation) when they are asked to join new initiatives.

The Balanced Scorecard and Key Performance Indicators

To translate vision and strategy into objectives and measurable goals, many companies use a balanced scorecard (Kaplan and Norton 1996). This dashboard helps managers keep their finger on the pulse of the business. The original balanced scorecard featured four broad categories: (i) financial performance; (ii) customers (e.g. customer satisfaction); (iii) internal processes (e.g. efficiency, safety); and (iv) learning and growth (e.g. morale), and aimed to balance (often short-term) financial and longer-term nonfinancial performance by providing a broad view of the business. Typically, each category includes two to five key performance indicators (KPIs), customized by each organization, based on its strategy, and implemented in its own dashboard.

The goal is to derive a set of measures matched to the business so that performance can be monitored and evaluated. If the business strategy is to increase market share and reduce operating costs, the indicators may well include market share and cost per unit. A business that emphasizes financial indicators like price and margin may disregard market share for higher-priced niche products. A list of objectives, targets, measurements, and initiatives comes with each indicator. The saying "we manage what we measure" holds true.

Company-wide KPIs break down into business unit, department, even work-team KPIs. Finally, we distinguish between lagging, real-time, and lead indicators of:

- past performance (lag indicators)
- current performance (real-time indicators)
- future performance (lead indicators).

Lag indicators include traditional accounting indicators such as: profitability, sales, and shareholder value; customer satisfaction; product and/or service quality; and employee satisfaction (Kenett and Salini 2011). They are useful, because they indicate the overall health of the business. At the same time, some liken using them to steering a car by looking in the rearview mirror.

Real-time indicators help determine the current status of a project. Cost performance index (CPI), for example, indicates the current status of funds expended. It measures budgeted cost of work against actual cost of work performed. A ratio less than one indicates that the project is overrunning its budget. Schedule performance index (SPI), as another example, indicates the current schedule status. It compares budgeted cost of work performed (BCWP) to budgeted cost of work scheduled (BCWS). A ratio less than one indicates that the work is behind schedule. For examples see Kenett and Baker (2010).

Lead indicators, as opposed to lag and current indicators, are designed to predict future performance. They are derived from:

- customer analyses (segments, motivations, unmet needs);
- competitor analyses (identity, strategic groups, performance, image, objectives, strategies, culture, cost, structure, strengths, and weaknesses);
- market analysis (size, projected growth, entry barriers, cost structure, distribution systems, trends, key success factors);
- environmental analyses (technological, governmental, economic, cultural, demographic, etc.).

Understanding the organizational vision, its key strategic initiatives, and the indicators used to run the company is basic if data scientists want to be effective. They should also build their networks and take a larger look at the company through their work, what we call "the data lens."

The Data Lens

As discussed above, one can look into an organization in a variety of ways: via its income statement and balance sheet; via the leadership team; via its business plan; and so forth. These projections are available to everyone.

The so-called "data and information lens" is uniquely available to data scientists, and it provides an extremely powerful end-to-end look. To employ it, one first examines the movement and management of data and information as they wind their way across the organization. This lens reveals who touches them, how people and processes use them to add value, how they change, the politics surrounding seemingly mundane issues like "data sharing," how the data comes to be fouled up, what happens when the data is wrong, and so forth. This is a great way to understand many of the problems and opportunities facing a company today.

Build Your Network

In Chapter 2, we noted that great data scientists have enormous networks. Establishing a network of sufficient breadth and depth takes time, patience, and initiative. Obviously some people are better at it than others. On the other hand, a data scientist, isolated from the organization, is almost certainly ineffective. We emphasize here the human element of data science – human beings are responsible for the annual report, SWOT, balanced scorecard, and KPIs. You should get to know them. And you should validate what you see through the data lens by talking to various people in different positions.

One final point: the best way to build a great network is to be part of other people's networks. Be generous in helping others understand data science.

Implications

If you are going to help people make better decisions, you need to understand them and the context in which they make decisions. This means immersing yourself in the business. One way a data scientist can do this is to be embedded with them. Hahn (2003) coined the term "embedded statistician" when he worked at General Electric. Of course, not all data scientists are embedded (more on the best organizational spots for data scientists in Chapter 15), but they should act as though they are. Thus, the real work of data scientists involves learning all they can about "the business," how it really operates, the values of both the organization and people working there, and who makes the really important decisions.

4

Understand the Real Problem

A Telling Example

One day, one of us (Redman) was approached by a middle manager seeking a sample size calculation. Something about performance in his group bothered him and he wanted to do a study to learn more. He needed extra budget to do the study, and he needed the sample size to justify that request. He confessed that he had visited another statistician earlier. The meeting had gone reasonably well, but that statistician had given him a formula, which he didn't really understand. He just wanted "the number."

Seem straightforward enough? The manager has explained the context (i.e. his budget request), his specific question (i.e. sample size), and his level of sophistication (i.e. can't use the formula). Plenty on which to proceed.

But Redman probed further:

- What was bothering him?
- What sort of study did he plan to conduct?
- How would he analyze the data once he collected it?

A wide-open exchange followed, covering these topics and more. It emerged that the manager did not know enough to design a full-blown study. Instead, he should "turn over some rocks to see what crawls out."[1] The two discussed a semistructured method for doing so. They also agreed that this method might not work, so they should talk frequently.

The discussion on how many rocks to turn over went like this:

MANAGER:	"So back to our original question. What's my sample size?"
REDMAN:	"If I told you 50, could you do that without additional resources in the next two weeks?
MANAGER:	"Hmmm. Probably not."
REDMAN:	"OK, what if I told you 25?"

---

[1] George Box (1980) noted "that if you don't know anything, no experiment could be designed." We interpret that observation as "turn over some rocks to see what crawls out" as the first step in remedying the situation.

The Real Work of Data Science: Turning Data into Information, Better Decisions, and Stronger Organizations, First Edition. Ron S. Kenett and Thomas C. Redman.
© 2019 Ron S. Kenett and Thomas C. Redman. Published 2019 by John Wiley & Sons Ltd.
Companion website: www.wiley.com/go/kenett-redman/datascience

MANAGER: "I'm not certain. Maybe."
REDMAN: "OK, your number is 25. Let's talk again in two weeks and we'll figure out what to do next."

To complete the story, the manager came back two weeks later, having completed 20 (of the 25). In 19, he had found a significant error. Some errors looked similar (the manager did not yet understand "common cause"). He concluded that there was little doubt that the "overall process was broken." Subsequent work should focus on understanding the larger implications for his division (a far larger organization than the group he managed) and sorting out what to do about them.

Understanding the Real Problem

Note the contrast between the original need and context and those arrived at during the first meeting. This example provides an object lesson in understanding the real problem. Data scientists simply must engage with "customers" in their languages and talk through the apparent problems to discover the real ones.

Frankly, with two warnings, we find that many people make this more difficult than it needs to be. Three points to think through. First is our choice of the word "customer." We find that viewing decision-makers as customers humanizes them. A customer can come from a different part of the organization, be way up in the hierarchy, or be based in another part of the world. But customers are people, with strengths, weaknesses, hopes, and fears. They are just like us in that respect.

Second, "in their languages." Just as we don't want to get into the technical specifications about how our car windshields were manufactured, so too we should not expect customers to engage with us in the (to them) arcane languages of statistics and data science. We should encourage them to speak their languages and make every effort to learn that language. This is hard work – after all, those who drill for oil, run social media campaigns, and hedge securities have developed their own specialized languages that are no more familiar to us than data science is to them. So don't be afraid to ask questions and to say things like "let me make sure I understand."

Third is "discovering the real problems." It is a rare decision-maker indeed who can articulate the real problem in the first try. Articulating problems is hard work, and there are so many special cases, external factors, and political considerations that may cloud one's head. It is the data scientist's job to help clear away the clutter, draw your customer out, and propose various options. It is tempting to rush this work, but don't. After all, as Albert Einstein observed, "if I had only one hour to save the world, I would spend the first fifty-five minutes defining the problem, and only five minutes finding the solution."[2]

The first warning involves bad intentions. Some decision-makers are not above using data science to justify decisions they've already made (more on this in Chapter 11), to made others look bad, or to promote personal agendas. So it also helps to develop a keen sense of smell – if something smells bad, it probably is. You are unlikely to fully prevent such behavior, but you must inform your boss and proceed with caution.

---

[2]http://www.azquotes.com/quote/811850.

The second warning involves scope creep. The discussion with the manager in the example above could just as easily have gone along the following lines:

sample size → general lack of management data → a need for new business
intelligence (BI) tools → a culture that doesn't invest in technology

All of these might be real, even important, problems, but they quickly grow far beyond scope. And beware, there are always forces that will complicate even the simplest problem.

To guard against scope creep, especially early on, we recommend that you limit yourself to problems that can be solved quickly, using existing resources, and within existing budgets. Redman and the manager in the example above did just that. After they solved the first problem, they defined the next problem, and the next. In doing so, they took on larger, more complex issues. But they did so guided by hard facts, experience, and increasing confidence.

There are, of course, problems that require you to think longer term and spend real money from the very beginning. Developing a predictive model that optimizes profits from credit decisions or works out the proper dosage for a new cancer medication are good examples. But the same thinking applies.

Understanding the problem is the first step in the life cycle we introduced in Chapter 1. Here the domain expert ecosystem is translated into the analytic ecosystem, based on our understanding of the problem. A poor translation can have disastrous results. As an example, Schmarzo (2017) describes the negative effects of the Medicare Access and CHIP Reauthorization Act of 2015, or MACRA. Among the major provisions of MACRA is the Quality Payment Program. Under the Quality Payment Program, physicians and nurses receive positive, neutral, or negative Medicare payment adjustments based upon a "Patient Satisfaction Score." But satisfying patients and helping them get better are not always the same thing, and the program had a negative consequence on patient outcomes.

On a more humorous side, Box (2001) told the story of a man who was very tall and his 4-year-old son. They were walking to get a newspaper, and the father suddenly realized that the little boy had to run to keep up with him. So he said, "Sorry, Tommy am I walking too fast?" And the little boy said, "No, Daddy. I am."

More examples on good problem solicitation as a prerequisite to proper statistical analysis are provided in Kenett and Thyregod (2006).

Implications

A recurring theme throughout our careers has been how often managers who admit that something "just doesn't feel right" have been spot-on. There are few real facts, and the initial problem is to get some. We call these "We don't know anything and need to sort out what's going on" problems. They occupy one end of a continuum of problems on which data scientists should engage. On the other end of the continuum are what we call "optimality problems." Current efforts work tolerably well, and the problem is to optimize performance or save money. And some problems occupy the middle ranges of the continuum.

Thus, the real work of data scientists involves listening to decision-makers, learning their languages, translating their languages into yours and back, engaging in ways that make them feel comfortable, and working together to sort out real problems that they can address. Get very good at this.

5

Get Out There

Lawrence Peter (Yogi) Berra once observed, "You can learn a lot just by watching." It is especially poignant for data scientists who must "get out there" and learn all they can about everything surrounding the data they use.

There are nuances and quality issues in the data you simply cannot understand sitting at your desk. Further, the world is filled with "soft data," relevant sights, sounds, smells, tastes, and textures that are yet to be digitized – and may never be. Things like the electricity in the air at a political rally, the smell on a cancer patient's breath, and the fear in the eyes of an executive faced with an unexpected threat. Just as data scientists must understand the larger context, the real problems, and the opportunities, as discussed in the previous two chapters, so too they must understand how the data they analyze was collected, in great detail.

Of course, the importance of these efforts is nothing new, as Joiner (1985, 1994), Hahn (2007), Hahn and Doganaksoy (2011), Kenett (2015), and Kenett and Thyregod (2006) attest.

Understand Context and Soft Data

This section is about contextual background and soft data. Great data scientists know that the only way to acquire this smorgasbord of information is to go get it. So they spend time on the road with truckers, probe decision-makers, wander the factory floor, pretend to be a customer, visit call centers, ask experts in other disciplines for help, and so forth. They delve deeply into processes of data creation and the complexities of measurement equipment. They ask old hands how their recommendations will be used, the likely results, and what can go wrong.

We have already explored how failing to get out there contributed to the poor showing of pollsters in the 2016 presidential election in the United States (Chapter 2).

In the 1980s, one of us (Kenett) was director of statistical methods at Tadiran, a large telecommunications corporation. He was appointed to this job after his CEO attended a Deming seminar. In these seminars, Deming recommended that organizations seeking to gain competitive advantage, improve quality, and increase productivity create such a position (more on this in Chapter 15). In that position, Kenett was able to drive process improvement initiatives and applications of statistical methods such as designed experiments and statistical process control. Tadiran became an industry leader with innovative products and high-efficiency processes.

The Real Work of Data Science: Turning Data into Information, Better Decisions, and Stronger Organizations,
First Edition. Ron S. Kenett and Thomas C. Redman.
© 2019 Ron S. Kenett and Thomas C. Redman. Published 2019 by John Wiley & Sons Ltd.
Companion website: www.wiley.com/go/kenett-redman/datascience

A Visit to the Production Floor

Kenett (left), as director of statistical methods of Tadiran Telecom, explains to the CEO of the Israel Aircraft Industry how process control and designed experiments helped reduce solder defects from 30,000ppm to 15ppm with significant savings and increased quality.

Many managers visited the company's production floor and development labs to get ideas on what can be achieved, a sort of benchmark (see box).

As another example, consider the oil business. Where the oil is thick, it is hard to pump out of the ground. To make this process easier, companies heat the oil first using steam. Steam is expensive and must be used according to strict ecological guidelines, so putting the right amount in is critical. There are many factors to consider – the underlying geology, the current temperature of the oil, the well's production history, and so forth – in working out the optimal amount of steam. All this can be done in front of a computer.

Data scientists seeking to understand the full context would also spend some time in the oil field. There, they would notice that the probe used to estimate current temperature is sometimes lowered into the well clean, while at other times it is covered with mud. As it happens, mud is a great insulator, leading to a "too-low" temperature and, in turn, too much steam. Having verified this through a simple experiment, the data scientist can now tackle the root of the issue, namely, the lack of a work instruction advising the technician to insert a clean probe. In this case, optimizing the amount of steam is important, but rooting out the data quality issue (the mud-covered probe) is more fundamental and saves millions. It illustrates a side benefit to getting out there – namely, identifying opportunities that others don't.

Not every data scientist spends enough time understanding these deeper realities data scientists study. Some are uncomfortable dealing with others and concentrate too much on "the numbers." It is especially important to **see how the data is actually collected,** because so much can go wrong. Measurement instruments get clogged with sand, pollsters do not follow their scripts, and survey developers inadvertently design their instruments in ways that bias results (Surveytown 2016). You can't simply assume that your data is unbiased and correct. You must sort out the nonsampling error and measurement variation. Finally, you must make sure all the data hangs together. In a factory, this means that individual parts should be traceable to the work order, measurements should be traceable to the measuring device, and the calibration history of the measuring device should be retrievable. Take a hard look, in person.

Identify Sources of Variability

Seeing actual data collection yields another important benefit as well – it helps the data scientist develop a better sense of the sources of variability. Domain experts, including engineers, knowledge workers, and service employees, can be helpful, but they are not accustomed to thinking about variation, whereas this is a strength of good data scientists.

So work hard to understand their perspectives, but don't depend on them. In particular, note suspected sources, their nature, why you expect they cause variation, and their dynamics. Then look for them in your data analyses, especially seeking to understand how to control or represent them.

A great example of why understanding variation in data inputs is so important is illustrated in a sketch on the BBC Scotland *Burnsitoun* comedy show (BBC 2017). Two Scots are in a voice-activated elevator that does not understand their accents. Obviously, not everyone speaks the Queen's English and the system was not designed to handle variability in the spoken word. Predictive models build using AI must also address variability in inputs. Understanding variation is key to predictive maintenance in advanced manufacturing (Barkai 2018).

Selective Attention

Now a couple of words of caution and possible opportunity. First, beware "selective attention," where one pays too much attention to certain details, while missing the larger context. In a popular video, two teams of three players pass basketballs to one another. One team has white shorts, the other has black shirts. People are asked to carefully watch the video (which is 1.21 minutes long) and count the number of passes by the black team (Simon 2010). Typical counts range from 9 to 15, which itself calls into question people's abilities to count events just by watching.

The punch line is different, however. Halfway through the video a black gorilla steps in, jerks its hand around, and leaves the frame. More than half of people seeing the video for the first time do not see the gorilla. In fact, when we rewind the video and point to the gorilla, many believe we tricked them. So keep your eyes open for the black gorillas wandering around that no one sees. The insulated probe in the oil field is just such an example!

Memory Bias

Second, beware memory bias as you talk to people. Elizabeth Loftus, a psychologist at the University of California, Irvine, studied the forces that taint people's memories after an experience is over, and she has consulted on hundreds of criminal cases. "Just because somebody tells you something, and they say it with a lot of detail and a lot of confidence and a lot of emotion, doesn't mean it really happened," she says (Loftus 2013; Baggaley 2017).

Much soft data resides in the memories of employees and may be distorted. And people often have their own agendas and biases, further distorting their memories. So talk to people, but be cautious.

Implications

So much of the nuance in the real world is not properly captured in the data, the metadata, or management reports. Data scientists need to understand this nuance. Engaging in water cooler discussions and talking to experts helps, but it has limitations. Touring facilities, visiting customers, joining a service technician on one of his or her shifts, and riding with a trucker provide insights that cannot be gained any other way. At a minimum, the soft data gathered in this way provides a context for the numbers you analyze. And often so much more.

Thus, the real work of a data scientist involves talking to people, visiting the places where the data is created, and delving into the details.

6

Sorry, but You Can't Trust the Data

This chapter is about data quality, which is a plague on data scientists and CAOs (and practically everyone else for that matter). It takes up to 80% of data scientists' time (Wilder-James 2016) and is the problem they complain about most (Kaggle 2017). Worse, you can never be sure you've found all the errors and, worse still, the issues grow larger and more impactful with AI. The data needed for data science can come from primary sources (created with similar objectives) or secondary sources (someone else collected it, often for a different objective, and processed it before it reaches you).

While there is no panacea, we can:

1. help you understand the extent of the problem;
2. provide some structure for dealing with immediate issues;
3. advise you to push your company get in front of recurring data quality issues.

Most Data Is Untrustworthy

Without delving too deeply into details, to be judged of high quality, data must meet three distinct criteria (Redman 2016):

- It must be "right:" correct, properly labeled, de-deduped, and so forth.
- It must be "the right data:" unbiased, comprehensive, relevant to the task at hand.
- It must be "(re)presented in the right way." For example, people can't read bar codes, locally used acronyms may confuse others, and so forth.

Regarding the first criteria, the most comprehensive study of data quality statistics that we know of was conducted in Ireland in 2014–2016 (Nagle et al. 2017). It made use of the "Friday Afternoon Measurement" (summarized in the next section), focused on the most important and recent data, and concluded the following:

- On average, 47% of newly created data records have at least one critical (e.g. work-impacting) error.
- Only 3% of the data quality evaluations can be rated "acceptable" using the loosest-possible standard.

The Real Work of Data Science: Turning Data into Information, Better Decisions, and Stronger Organizations,
First Edition. Ron S. Kenett and Thomas C. Redman.
© 2019 Ron S. Kenett and Thomas C. Redman. Published 2019 by John Wiley & Sons Ltd.
Companion website: www.wiley.com/go/kenett-redman/datascience

- The variation in data quality is enormous, with individual data quality evaluations on a 0–100% scale, ranging from 0 to 99%. Still, deeper analyses revealed no important industry (e.g. health care, tech), data type (e.g. people data, customer data), organization size, or public/private differences.

Thus, no sector, government agency, or department is immune to the ravages of extremely poor data quality. And importantly, these results do not include issues such as duplicates, inconsistencies between systems, and so forth. Further, they focus only on the most recent and important data that is under an organization's control. They do not include older or less-used data or data not under the organization's control. Thus, bad as they are, these results reflect an upper bound on what you have to deal with in data science.

Data Quality and the Internet of Things

Those who deal with automated measurement (e.g. the Internet of Things (IoT]) are sometimes tempted to dismiss these results, thinking they stem from human error. Doing so is unwise. First, while some errors are human related, most are not. Second, our experience, although anecdotal, convinces us that automated measurement is no better, although the failure modes may be different. For example, a meter in the electric grid may simply shut down, or sand may clog an anemometer and cause intermittent failures. So until a specific device is proven correct, you should assume it produces data of no higher quality.

"The right data" standards are murkier. After all, what is just right for one analysis may not suit another. Data scientists usually consider anything they can get their hands on, but that may not be good enough, as reports on bias in data used for facial recognition (Lohr 2018) and criminal justice (Tashea 2017) attest.

We don't know of any definitive study on data presentation either. Still, issues come up from time to time. For example, handwritten notes and local acronyms have complicated IBM's efforts to apply AI (e.g. Watson) to cancer treatment (Ross 2017).

Importantly, increasingly complex problems demand not just more data but more diverse, comprehensive data, and with this come more quality problems. For example, dealing with subtle differences in the definitions of data from different sources is increasingly challenging.

Of course, the caustic observation "garbage in, garbage out" has plagued analytics and decision-making for generations. The concern today is "big garbage in, big garbage out." Data scientists bear special responsibility here; after all, the caliber of your recommendations depends on high-quality data!

AI and some predictive analyses exacerbate our concerns. Bad data can rear its ugly head twice – first in the historical data used in training a predictive model (PM) and second in the new data used by that PM going forward. Consider an organization seeking productivity gains with its machine learning efforts. Although the data science team that developed the PM may have done a solid job cleansing the training data, the PM will still be compromised by bad data going forward. Again, it takes people, lots of them, to find and correct the errors. This in turns subverts the hoped-for productivity gains.

Finally, there is the possibility of "cascades." A cascade occurs when a minor error in one prediction or decision grows larger in subsequent steps. The financial crisis that started in late 2007 is one example. Erred data in mortgage applications led to incorrect predictions of default rates. These, in turn, impacted the performance of securities (e.g. collateralized debt obligations) build on mortgages. And on and on. We are especially concerned that, as AI technologies penetrate organizations, the output of one model will feed the next, and the next, and so on, even crossing company boundaries.

It is tempting to ignore these issues, trust the data, and jump into your work. If the data falls short, you can always go back and deal with quality issues. After all, "innocent until proven guilty."

But both the "facts" (i.e. quality is low) and the "consequences" ("garbage in, garbage out") advise against it. And decision-makers are well aware of data quality issues. In a recent survey, only 16% agreed that they trust the data (*Harvard Business Review* 2013). Therefore, we recommend (and experienced data scientists know) that you adopt the position that "the data is not to be trusted, until proven otherwise."

Dealing with Immediate Issues

Not surprisingly, we recommend an all-out attack on data quality from the very beginning. This section focuses on dealing with immediate issues and the next on getting in front longer term.

The first step, as we discussed in detail in Chapter 5, is to visit the places where the data is created. There is so much going on that you can't understand in any other way.

Second, evaluate quality for yourself. If the data was created in accordance with a first-rate data quality program, you can trust it. Such programs feature clear accountabilities for managers to create data correctly, input controls, and make efforts to find and eliminate the root causes of error (Redman 2015). You won't have to opine whether the data is good – data quality statistics will tell you. You'll find a human being who will be happy to explain what you may expect and answer your questions. If the data quality stats look good and the conversation goes well, trust the data. Please note that this is the "gold standard" against which the other steps below should be calibrated.

You should also develop your own data quality statistics, the "Friday Afternoon Measurement" (Redman 2016), as used in the study noted above. Briefly, you lay out 10 or 15 important data elements for 100 data records on a spreadsheet (best if you do so for the 100 most recently created records). If the new data involves customer purchases, such data elements may include "customer name," "purchased item," and "price." Then work record by record, taking a hard look at each data element. The obvious errors will jump out at you – customer names will be misspelled, the purchased item will be an item you don't sell, the price may be missing. Mark these obvious errors with a red pen. Then simply count up the fraction of records with no errors. In many cases you'll see quite a bit of red – don't trust this data! If you see only a little red – say, less than 5% of records with an obvious error – you can use this data with caution.

Look, too, at patterns of the errors. If, for example, there are 25 total errors, 24 of which occur in the price, eliminate that data element going forward. But if the rest of the data looks pretty good, use it with caution.

Third, work through the "rinse, wash, scrub" cycles. "Rinse" replaces obvious errors with "missing value" or corrects them if doing so is very easy; "scrub" involves deep study, even making corrections one at a time, by hand, if necessary; and "wash" occupies a middle ground.

Even if time is short, scrub a small random sample (say, 1,000 records), making it as pristine as you possibly can. Your goal is to arrive at a sample of data you know you can trust. Employ all possible means of scrubbing and be ruthless! Eliminate erred data records and data elements that you cannot correct, and mark data as "uncertain" when applicable.

When you are done, take a hard look. When the scrubbing has gone really well (and you'll know it if it does), you've created a data set that rates high on the trustworthy scale. It's OK to move forward using this data. Sometimes the scrubbing is less satisfying. If you've done the best you can, but still feel uncertain, put this data in the "use with caution" category. If the scrubbing goes poorly – for example, too many prices just look wrong, and you can't make corrections – you must rate this data, and all like it, as untrustworthy. The sample strongly suggests none of the data should be used in any analyses, going forward.

After the initial scrub, move on to the second cleaning exercise: washing the remaining data that was not in the scrubbing sample. Because scrubbing can be a time-consuming, manual process, the wash allows you to make corrections using more automatic processes. For example, one wash technique involves "imputing" missing values by statistical means (see Wikipedia 2018a). Multiple imputation is a statistical technique for analyzing incomplete data sets, that is, data sets for which some entries are missing. Application of the technique requires three steps: imputation, analysis, and pooling (Rubin 1987), An up-to-date account of multiple imputation, as well as code and examples using the mice package in R, can be found in van Buuren (2012). If the washing goes well, put this data into the "use with caution" category.

The following flow chart (Figure 6.1) summarizes this logic.

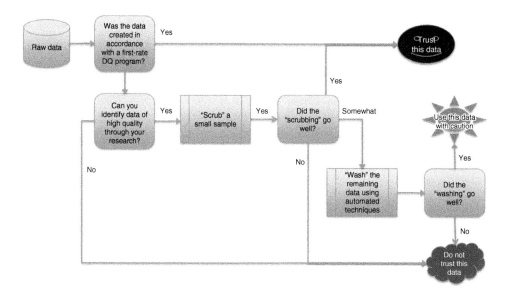

Figure 6.1   Process for evaluating data's trustworthiness. DQ: data quality.

Once you've identified the data that you can trust or use with caution, integrate disparate data sources. You need to do three things:

- Identification: verify that the Courtney Smith in one data set is the same Courtney Smith in others.
- Alignment of units of measure and data definitions: make sure Courtney's purchases and prices paid, expressed in "pallets" and "dollars" in one set, are aligned with "units" and "euros" in another.
- De-duplication: check that the Courtney Smith record does not appear multiple times in different ways (say, as C. Smith or Courtney E. Smith).

At this point, you're ready to do your analyses. When possible, conduct analyses using the "trusted data" and the "use with caution data" in parallel. Pay particular attention when you get different results based on "use with caution" and "trusted" data. Both great insights and great traps lie here. When a result looks intriguing, isolate the data and repeat the steps above, making more detailed measurements, scrubbing the data, and improving wash routines. As you do so, develop a feel for how deeply you should trust this data. Please note that this comparison is not possible if the only data you truly trust is the scrubbed 1,000-record sample and you're using AI. One thousand trusted records is simply not enough to train a PM.

Maintain an audit trail, from original data, to steps you take to deal with quality issues, to the data you use in final analyses. This is simply good practice, although we find that some skip it.

Understanding where you can trust the data allows you to push that data to its limits. Data doesn't have to be perfect to yield new insights, but you must exercise caution by understanding where the flaws lie, working around errors, cleaning them up, and backing off when the data simply isn't good enough.

No matter what, don't get overconfident. Even with all these steps, the data will not be perfect, as cleansing neither detects nor corrects all the errors. Finally, be transparent about the weaknesses in the data and how these might impact your analyses

Getting in Front of Tomorrow's Data Quality Issues

Of course, today's data quality problems are bad enough. Even worse is failing to take steps so they don't recur. If you're experiencing a 20% error rate now, you can feel confident that you will experience a 20% error rate in the future (within statistical limits, of course). And growing data volumes mean more data errors, and more cleanup, in the future.

Worst of all, bad data affects everything your entire organization does. After all, much of the data you use in analytics is used by others in basic operations. For example, an incorrect address may slow one of your analyses, but it also means that someone's package was not delivered, on time or at all. While there is considerable company-to-company variation, best estimates are the data quality costs a typical organization 20% of revenue (Redman 2017c).

The only solution is to find and eliminate root causes of error (Redman 2016). Leading efforts to attack data quality across the entire company is beyond the scope for most data scientists today. Still, data scientists may have the best, widest view of data quality issues. Summarizing them, including their costs and what should be done about them, though, must be in the data scientists' and CAOs' wheelhouses!

Validation sampling
1. A subset of quarterly discharge medical records, originally abstracted by the primary data collection staff, for a given measure should be sampled for reabstraction by a second staff responsible for data validation.

   - **Approximately 5% of the abstracted records should be targeted for reabstraction for a given measure in a given quarter.**
   - The minimum quarterly sampling requirement for reabstraction is nine sampled cases per measure.
   - If the originally abstracted quarterly medical record size is less than 180 cases, then the minimum sample requirement for reabstraction would be nine cases.

Figure 6.2   JCI data validation guidelines.

This admonition applies for entire industries as well. Here standards can help. For example, the Joint Commission International (JCI), which accredits hospital procedures all over the world, has devised guidelines to ensure data quality. Some involve definitions, such as what qualifies as an "infection" in counts of infection. Others involve controls, such as a data validation step whereby two independent and qualified people retrieve data from a hospital's system and compare results. Figure 6.2 is an extract from the data validation guidelines (Joint Commission International 2018).

Implications

Data quality is probably the toughest issue data scientists face. Worse, it impacts your entire organization. Thus, the real work of data scientists involves stepping up to the near-term issues and addressing them in a coordinated, professional manner. And the real work of CAOs involves clarifying the larger issues for the rest of the company and helping start programs that get to the root causes of these issues.

7

Make It Easy for People to Understand Your Insights

Most of us learn more from our mistakes than we do from our successes. One of us, Redman, learned a lesson that has stuck with him for 30 years.* The story concerns his first big presentation at AT&T headquarters. He completed his preparation well in advance and rehearsed carefully. Then off to the big meeting.

It could not have gone worse! The only impressions he left were bad ones. Young hothead that he was, he blamed everyone but himself, including the audience: "The average manager up here can't even understand a pie chart!"

An established veteran of many such presentations looked him square in the eye and said, "Of course not, Tom. It's your job to make it so they don't have to."

The lesson is not simply about Redman being immature. As a data scientist, you face a tall order in getting decision-makers to comprehend and believe data, your results, and their implications. You have to think through their background and present in ways that advance their understanding. This takes more time than you may think. It helps to be a good writer and speaker. But more than anything, great visuals (e.g. graphics) and great stories carry the day! Visualization has a storied past – see Fienberg (1979). Communication is the eighth dimension in the information quality framework introduced in Chapter 13; see also Kenett and Shmueli (2014, 2016a) and Appendix C.

First, Get the Basics Right

At a minimum, you must make your plots and the accompanying explanations easy to understand. As Edward Tufte advises (Tufte 1997), clearly label the axes, keep chart junk to a minimum, and don't distort the data.

Consider this example. The plot in Figure 7.1 is a typical result of a well-conceived and well-executed data quality program. But it features too many unfamiliar terms, such as "accuracy rate" and "fraction perfect records." Without additional explanation, you'll lose your audience.

---

*This Chapter is based, in part, on a *Harvard Business Review* digital article by Redman (2014).

The Real Work of Data Science: Turning Data into Information, Better Decisions, and Stronger Organizations, First Edition. Ron S. Kenett and Thomas C. Redman.
© 2019 Ron S. Kenett and Thomas C. Redman. Published 2019 by John Wiley & Sons Ltd.
Companion website: www.wiley.com/go/kenett-redman/datascience

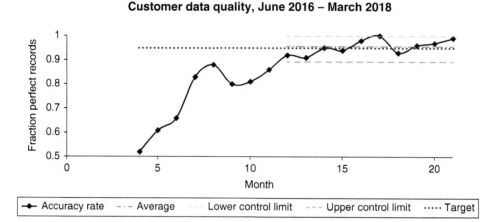

Figure 7.1 The plot of data quality results, as first presented (second-year average and target are superimposed).

So, start by explaining how to interpret the chart at its most basic level: "Here is a time-series plot of the results of our data quality program. I know most of you are familiar with such plots, but let's make sure we're all on the same page here. As you can see, we focused on the quality of customer data. The x-axis is time, and here I am showing one point every month. The y-axis is the fraction of data records that were created perfectly each month. That's how we are measuring accuracy. It is a tall standard and I shall have more to say about that in a minute." Then, explain to your audience how to read the data presented within the chart: "The solid line with diamonds displays our actual results. The dotted line shows the target we set for ourselves, and the dashed lines are control limits around the second-year average (the dashed–dotted line). These are a bit technical and explained later. Now before we dig in, are there any questions about how to read the chart?"

Note that you have told your audience where you will be expanding, but you are spending these early moments focusing on the basics of reading the chart first. This lets them fully comprehend the visual, so they can then put their full attention toward listening to your explanation of the data to come.

Now tell the story of the data in a powerful, animated fashion. In this case, there is much to tell, including how and why the program started; the joys and challenges surrounding the documentation of customer requirements; measurements against those requirements, including the logic of the choice of metric on the y-axis; improvement projects; and how you established control – essentially the implications of those dashed lines. Point out the impact of each on the plot as you proceed. Use Figure 7.2 instead of Figure 7.1.

Different audiences will have different needs, and you should tell the story in the simplest and most direct way that you can for each one. For example, a technical community may wish to understand the details in your choice of metric and the software used to draw the plots. A senior decision-maker may wish to understand the significance of the story for extending data science across the organization. While the main story will be the same for each, the emphases should be very different.

As we have already noted, many people are skeptical about analytics, AI, big data, data science, and statistics (many recall Twain's observation that "There are several kinds of lies: lies, damned

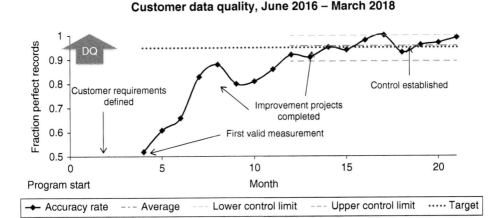

Figure 7.2   The earlier data quality plot, fully explained. Note the addition of the "this way is good" arrow. This addition is particularly helpful for those who read the plot, rather than listen to the accompanying story (as in Figure 7.1, second-year average and target are superimposed).

lies, and statistics").[1] Whether this skepticism is justified or not, it does enormous damage by slowing, or even stopping, the penetration of good ideas into organizations. As a data scientist, you have a sacred trust to build support with decision-makers. You must do the following:

1. Present the facts in the most straightforward, accurate way you can. This is especially true when the results are not favorable. Further, if your results are counter to established wisdom, simply state that this is the case.
2. Present a comprehensive picture. Leaving out a key fact is the worst kind of lie.
3. Provide proper context, including where the data originates and what you have done to ensure it is of high quality. (And if you have done little, you must explicitly state, "The data is of unknown quality. This could impact results.")
4. Summarize your analysis, including shortcomings and alternative explanations for the results you see.

It is fine (and often appropriate) to state your opinion, but you must clearly separate your opinion from the facts.

Presentations Get Passed Around

One more critical point. Successful oral presentations live on as people pass on PowerPoint decks or links to them. People reading a slide deck alone will not have the benefit of your oral explanations, so you must think of their needs as well. An old saying from Bell Labs advises, "People spend an average of 15 seconds looking at a chart. Don't make them spend 13 of those seconds figuring out how to read the chart. Build in explanations wherever possible. Even better, make the graphic tell the story."

---

[1]There is actually a different way to read this statement misattributed to Twain. If the second comma is considered a hanging comma, statistics is put in contrast to "lies and damn lies."

With this in mind, take two steps. First, provide your explanation of how to read the chart in the notes page of your PowerPoint or slide deck program. Second, annotate the graphic, as done in Figure 7.2. Although annotations do not replace a well-told story, they do give the reader some inkling of what is involved.

The Best of the Best

Finally, to most decision-makers, an ounce of insight is worth a ton of analysis. Thus, one outstanding graphic that cuts to the heart of the issue at hand and guides next steps is worth more than hundreds of mediocre ones. Seek that graphic. Presented this way, data science is raw power!

There are dozens of books on this topic, and all good analysis packages help you develop amazing graphics. We've already cited Tufte. Other examples follow.

Hans Rosling (2007) developed bubble plots to show how phenomena evolve over time. He claims that preconceived ideas inhibited academics from properly realizing trends in demographics and child mortality statistics. His presentations showed that, in this respect anyway, third-world countries are not third world at all.

Jean-Luc Doumont (2013) trains scientists to communicate more effectively. Among other things, he highlights the pitfalls in communication with PowerPoint.

Jorge Camoes (2017) provides 12 suggestions for effective thinking with data visuals, with examples. His main points are the following:

- Data visualization is not enough; you have to have the contextual knowledge to detect and interpret patterns.
- Visualization fails not because there are too many data points but because the presenter doesn't understand the data or does not care about the message.
- Simplicity is not minimalism or removing junk. Not surprisingly, he advocates removing the irrelevant, minimizing the accessory, adjusting the necessary, and adding the useful.

Implications

The practical reality is that few decision-makers understand, or even care about, p-values, significance tests, and the like. Nor do they appreciate how hard you work to tease insights from the data. But they do care about results, particularly those presented in simple, visual, compelling ways that are directly applicable to the problem at hand. Thus, the real work of data scientists involves telling compelling stories based on your insights, providing simple graphics that drive the message home, and helping decision-makers understand the full implications and context. Data scientists need to get very good at this.

8

When the Data Leaves Off and Your Intuition Takes Over

It may seem trite to mention, but the power in data science lies not so much in uncovering what the data reveals about itself but what it reveals beyond itself. In this chapter, we focus on the process of moving from numbers to data, to information and insights (Kenett 2008), what we call "generalizability" in the information quality framework to be introduced in Chapter 13.

We recognize that some data is "soft," based on a person's experience, impressions, and feelings. Soft data stands in contrast to "hard data," particularly that which is digitized. And although we often prefer hard data, data scientists must not discount soft data. It is often valid and may well go far beyond the hard data. Good data scientists aim to combine the two. See Appendix B for a full discussion of what we mean by hard data, soft data, and information.

Figure 8.1 summarizes the situation. We want to make valid inferences and predictions from the data about bigger, more important areas of interest to decision-makers. Briefly, we recognize four distinct methods for doing so:

- The "laws of nature" refers to laws and models that allow one to extrapolate, under assumptions.
- "Statistical generalization" refers to making inference from a sample (of hard data) to a target population.
- "Domain-specific generalization" refers to applying domain knowledge, not fully supported by the hard data, to other circumstances, such as the future or different populations.
- "Intuition" refers to people's ability to reason from the data in ways that cannot be fully explained. Science in general and data science in particular often discount intuition. But it is undeniably true that some decision-makers have unerring intuition. And at the least, it is necessary because all decisions are made in the face of uncertainty.

To leave no doubt: data scientists should bring as much data, both hard and soft, and combine all types of generalizations in the most powerful, transparent ways they can to help decision-makers!

The Real Work of Data Science: Turning Data into Information, Better Decisions, and Stronger Organizations, First Edition. Ron S. Kenett and Thomas C. Redman.
© 2019 Ron S. Kenett and Thomas C. Redman. Published 2019 by John Wiley & Sons Ltd.
Companion website: www.wiley.com/go/kenett-redman/datascience

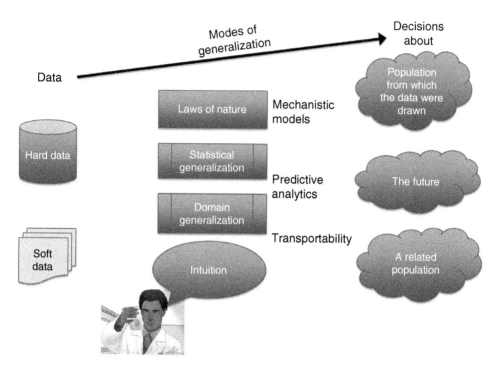

Figure 8.1   Modes of generalization.

Modes of Generalization

The best, most reliable form of generalization involves the laws of nature. These include conservation of mass, conservation of energy, conservation of momentum, Newton's laws, the principle of least action, the laws of thermodynamics, and Maxwell's equations. Sometimes these are called "mechanistic models of modes of action." These laws started as empirical laws that were embraced as laws of nature and have stood the test of time. They have been verified time and again, and today, we do not need further data to invoke them, only knowledge of physics, chemistry, biology, or other scientific disciplines

Mathematics, the queen of the sciences, offers a unique context. Paul Erdos, the famous mathematician, used to talk about The Book, in which God maintains the perfect proofs of mathematical theorems (Aigner and Ziegler 2000). The laws of nature build on The Book.

Now consider statistical generalizability. Sorting it out requires deep understanding of the goals (Chapter 4). In making inference about a population parameter from a sample, statistical generalizability and sampling bias are the focus, and the question of interest is, "What population does the sample represent?" (Rao 1985). In contrast, for predicting the values of new observations, the question is whether the analysis captures associations in the training data (i.e. the data used in model building) that generalize to the to-be-predicted situations. Control charts present a good example. The logic goes like this: "Assuming the process remains stable, we expect performance to vary within the upper and lower control limits. We further expect average performance to be close to the center line."

Statistical generalizability is commonly evaluated using measures of sampling bias and goodness of fit. In contrast, scientific generalizability, used for predicting new observations, is often evaluated by the accuracy of prediction of a hold-out set from the to-be-predicted population. This assessment is a crucial protection against overfitting, which occurs when your model fits previously collected data perfectly but does very poorly with new data.

Randomization lies at the heart of statistical generalization. As well as guarding against unknown biases, it provides the mathematical foundations that support calculation and interpretation of p-values, significance levels, and so forth. But there can be issues. Many decision-makers have a hard time understanding these concepts, just as many data scientists have a hard time explaining them.

Further, clinical trials may be subject to "sample selection-bias," because participation in a randomized trial cannot be mandated. Sample patients may consist of volunteers who respond to financial and medical incentives, leading to a distribution of outcomes in the study that differ substantially from the distribution of outcomes more generally. This sample selection bias is a major impediment in both the health and social sciences (Hartman et al. 2015). Data scientists must cope.

"Transportability" is another way to generalize. Transportability is defined as a transfer of causal effects learned in experimental studies to a new population, where only observational studies can be conducted. In a study on urban social interactions, Pearl and Bareinboim (2011, 2014) used transportability to predict results in New York City, based on a study conducted in Los Angeles, accounting for differences in the social landscape between the two cities.

Another example of generalization, in the context of personal ability testing, is the concept of specific objectivity (Rasch 1977). This testing is also known as "item response testing" (IRT). Specific objectivity is a theoretical state achieved if responses to a questionnaire, used to compare levels of students, are generalizable.

Yet another example, derived from online auction studies, provides one more example of the importance of precise clarification of the intent of generalization. In Chapter 1, we mentioned a study of the effect of reserve price on final price for eBay auctions as reported in Katkar and Reiley (2006). The authors designed an experiment that produced a representative sample of recorded Internet auctions. Their focus was on statistical generalization. In contrast, the study by Wang et al. (2008) forecasts new auction prices. The authors in Wang et al. (2008) evaluated predictive accuracy using a hold-out set instead of standard errors and sampling bias as used by Katkar and Reiley (2006). A third study, on consumer surplus in eBay, dealt with statistical generalizability by inferring from a sample to all eBay auctions. Because the sample was not drawn randomly from the population, Bapna et al. (2008) performed a special analysis, comparing their sample with a randomly drawn sample.

Domain-based (or scientific) expertise allows findings from specific data to be applied more generally (Kenett and Shmueli 2016a). Thus, marketing managers might base their decisions on how to run a marketing campaign in location A using a market study conducted in location B. They have no data on A, but their experience (soft data) tells them how to adopt the conditions in B to what is required in A.

Similarly, a software development manager, in the face of a limited testing budget, might decide to release a version with minimal testing because its functionality is basic, and the person who developed it has a good record. In other cases, the manager might decide to significantly increase the testing effort. Note that he or she made this decision without formal data analyses. The approach has benefits (i.e. speed) but carries risks that decision-makers should bear in mind.

Last, there is intuition. In the examples just above, both the marketing manager and the software development manager could explain the soft data and the rationales they used to reach their decisions. But even in the simplest, most straightforward situation, uncertainty remains. Thus, all important decisions are made in the face of uncertainty – if for no other reason than the future is unpredictable. And here intuition must take over!

Data scientists must embrace this reality, taking three steps: First, intuition should not replace sound inferences based on trusted data. Rather, it should take over where the data leaves off. They must help decision-makers understand the distinction. Second, wherever possible, they must quantify the uncertainty. And finally, they must cultivate their own intuitions.

Implications

Data scientists should view the various modes of inference as tools, just as they view R and Hadoop. And they should learn to use, and combine, them all. Too many data scientists prefer to stare at the data and do not think deeply enough about generalization of findings.

Thus, the real work of data scientists involves reasoning from the data to the situations of interest to the decision-maker. There are many ways to do so, and data scientists should embrace them all. Data scientists should strive to remove as much uncertainty as they can in their analyses and quantify (or at least clarify) that which remains. They must recognize that no analysis, no matter how thorough, removes all uncertainty, as there are just too many things that can go wrong or change. And they must develop their own intuition.

9

Take Accountability for Results

The wide-angle perspective of data science includes activities as diverse as building trust so you are asked to contribute to really important problems, clearly stating the problem, conducting the analyses, teaching, supporting decisions in practice, and so forth. This chapter focuses on one activity that is too often ignored – impact assessment.

Impact assessment is important so data scientists (and CAOs and groups of data scientists) can show others what they contribute in concrete terms. This in turn can help with funding, build trust, and more powerfully position the work. Similarly, it helps data scientists learn to become more effective. It is the last step in the life cycle introduced in Chapter 1.

Importantly, different communities judge impact differently. In science, new ideas must stand the test of time, and concepts such as statistical significance help guard against results that will not do so. In business, the criteria are wholly different. They may involve increasing sales, decreasing costs, improving market share, reducing risk, and so forth. Results, particularly those tested in the marketplace, need not stand the test of time, but they must stand up to tough competitors, gain new customers, and keep current ones. In a nonprofit organization, results may be judged on criteria such as improved test scores, reduced homelessness, enhanced national security, and the like.

In our view, none of these criteria are inherently better or worse; easier or more difficult; more noble or more basic than the others. But they are different, so understanding what the organization values is essential, as we discussed in Chapter 3.

Practical Statistical Efficiency

Of course, statisticians and others have understood the importance of evaluating the impact of their work for generations. Researchers of statistical methods have thought in terms of "statistical efficiency," comparing, for example, two methods for estimating the mean of a population. Building on a basic idea, Kenett et al. (2003) proposed the concept of practical statistical efficiency (PSE) to address the impact of statistical work in a specific problem area. PSE embraces the following elements:

$V\{D\}$ = value of the data actually collected
$V\{M\}$ = value of the statistical method employed (statistical efficiency)

The Real Work of Data Science: Turning Data into Information, Better Decisions, and Stronger Organizations, First Edition. Ron S. Kenett and Thomas C. Redman.
© 2019 Ron S. Kenett and Thomas C. Redman. Published 2019 by John Wiley & Sons Ltd.
Companion website: www.wiley.com/go/kenett-redman/datascience

$V\{P\}$ = value of the problem to be solved
$V\{PS\}$ = value of the problem actually solved
$P\{S\}$ = probability level the problem actually gets solved
$P\{I\}$ = probability level the solution is actually implemented
$T\{I\}$ = time the solution stays implemented
$E\{R\}$ = expected number of replications.

Let's explore each in turn.

$V\{D\}$ = Value of the Collected Data

The application of data science depends on data, so obtaining the right data of the right quality is critical. A high $V\{D\}$ corresponds to data being most relevant to the problem, trusted, clearly understood by relevant stakeholders, and collected comprehensively without bias. We discussed this in Chapter 6.

$V\{M\}$ = Value of the Analytic Methods Employed

This concept is closest to the original idea of mathematical statistical efficiency and includes the idea that the method should be as efficient as possible. As an example, suppose a manager wishes to reduce billing errors and must first obtain an accurate baseline error rate. Suppose there are two candidate methods, A and B. Method A is more efficient than method B if method A requires a smaller sample to provide the required estimate with the same prespecified error. More generally, a high $V\{M\}$ is assigned to methods with proven mathematical properties, such as unbiasedness and consistency.

$V\{P\}$ = Value of the Problem to Be Solved

Data scientists sometimes forget this part of the equation. Some might choose problems on the basis of technical depth rather than the value of solving them. To illustrate, one of us spent time figuring out how to reduce billing errors that were worth over $700,000/year, a fact crucial to management, even though solving the problem was not particularly difficult. A high $V\{P\}$ is assigned to problems of strategic importance to the organization.

$V\{PS\}$ = Value of the Problem Actually Solved

Usually no one method actually solves the entire problem, only part of it, so this part of the equation is expressed as a fraction of $V\{P\}$. In the case of the billing example, the manager expected to reduce the billings errors from 24,000 to 3,000 per billing cycle, a success rate of 87.5%. Problems with high $V\{P\}$ that are fully solved get a high $V\{PS\}$.

$P\{S\}$ = Probability the Problem Actually Gets Solved

This is both a statistical question and a management question. Did the method work and lead to a solution that worked, and were the data, information, and resources available to solve the problem? Part of this PSE component is related to management and technical personnel's buy-in and in meeting the challenge of facing the problem tackled. This is achieved by getting the relevant stakeholders to play an active role in specifying the problem and interpreting results. A high value of $P\{S\}$ implies that proper planning and effective execution have been carried out.

P{I} = Probability the Solution Is Actually Implemented
It is all well and good to propose grand solutions that look good in theory. But can they succeed in practice? And overcoming resistance to change is often the most difficult part of data science. A high value of P{I} implies that a proper match between management approach and analytic methods has been established. More on this in Chapter 16.

T{I} = Time the Solution Stays Implemented
Problems have the tendency to reoccur. This is why we emphasize holding the gains in any process improvement. Take the billing example – suppose that in the first year the company saves $700,000. With tight controls in place, the original problem stays solved, and the company saves more than $2 million in only three years. More generally, a high T{I} reflects a problem that stays solved for a long time.

E{R} = Expected Number of Replications
Good data science solutions are often replicable beyond their initial focus. A high E{R} represents a very large number of potential replications of the solution.

   A PSE assessment does not have to be a big, formal exercise – it can even be done qualitatively with a verbal description of the eight elements. One can also apply scores (say, 1–5) for each element and aggregate these using a multiplication formula or geometric means. What's important is structuring a full discussion of the various elements affecting the true impact of a specific project or collection of projects.

Using Data Science to Perform Impact Analysis

Further, one of the most important uses of data science involves helping others understand the impact of policy, legal, and methodological changes in both government and business. These can have enormous political overtones, but data scientists should not back away.

   An example of this is the assessment of changes conducted by the Australian Bureau of Statistics to make their longitudinal surveys less expensive. Part of the value of such surveys lies in generating time-series data, which makes them useful for social, economic, and environmental analyses and policy-making. Changing the survey methodology, platform, or questions can affect the continuity of the time series, making them harder to use in interpreting the impact of policy decisions, as one cannot know if changes are because of the policy or the new methodology. One approach to assess the impact of methodological changes is to run the old method and the new one in parallel for a time. Zhang et al. (2018) describe exactly how to do so.

   This parallel testing approach is essential because national statistical offices all over the world face the same efficiency improvement challenges. So do organizations carrying out customer or employee satisfaction surveys. It forms the basis of what is known as "A/B testing," which is widely used in web applications.

   In terms of the impact assessment, monetary savings are measured easily enough. But the Zhang et al. work is of great importance because it scores high on E{R}, the expected number of repetitions, across government and industry.

   Now consider as another example, the busing of school children, a topic as controversial as any. In one district of Tel Aviv, an experiment allowed some children to choose their middle and high schools, while others were not given this choice. Lavy (2010), a labor economist, evaluated the impact of this policy based on data on discipline, student social acclimation, student–teacher relationships, dropout rates, exam scores, and matriculation, under a strict privacy protection regimen. Data from adjacent cities, where the free choice option was not enforced, was used as a control.

Lavy's analyses showed that those who were given the choice of their schools showed an increased enrollment in academic colleges and a 5% increase in annual earnings at age 30. V{D}, the value of the data; V{PS}, the value of the problem solved; and T(I}, the time for which these benefits last, are all exceptionally high in the Lavy study.

Implications

Not surprisingly, most data scientists prefer to analyze data rather than to engage in an internal political debate. But the cold, brutal reality is that everything about data science is contentious – for example, there are plenty of smart, well-meaning people who believe data science is just the latest management fad. There are also plenty of others who believe that they will lose their power, position, even jobs if data science takes over. So, they fight it tooth and nail. Finally, many people trust their intuitions over data science findings, and they can rightly cite untrusted data as the rationale. Data scientists and CAOs ignore these realities at their peril!

Thus, the real work of data science includes selling it in a tough market filled with other good ideas, powerful special interests, and fear. Your best sales tool is solid results that advance the company, government agency, or nonprofit. This aspect of the work may prove uncomfortable for many. Our best advice is this: get over it!

10

What It Means to Be "Data-driven"

Over the past several years, the term *data-driven* has penetrated the business lexicon and appears to be here to stay.* *Data Driven* is the title of one of Redman's books, and academic work shows that companies that rate themselves as "data-driven" are measurably more profitable (McAfee and Brynjolffson 2012) than those that don't. Health care is undergoing a similar transformation, with evidence-based medicine becoming the worldwide standard for improved care delivery (Masic et al. 2008). So, becoming data-driven is clearly a worthwhile endeavor. But what exactly does it mean?

Data-driven Companies and People

A "data-driven" company is one that strives to make better decisions, by individuals and in decision-making groups, up and down the organization chart, every day. This means making slightly better decisions today than yesterday and slightly better decisions tomorrow than today. Forever. Moreover, one needs to recognize the time constraints associated with any given decision. After all, a good decision today may yield better results than a more informed one a month on. In the information quality (InfoQ) framework to be discussed in Chapter 13, this is referred to as "chronology of data and goal."

   This approach applies to individuals as well. There is considerable variation in the degree to which decision-makers embrace the concept.

   Another way of expressing what it means to be data-driven is as follows: all decisions are made in the face of uncertainty (see Figure 10.1). Suppose a decision-maker must make a decision today and, if he or she applies all the available hard data today, the uncertainty is reduced by 48%. Then intuition, perhaps trained by experience and soft data, must take over, so the decision-maker combines all three (hard data, soft data, and intuition) to make the best decision he or she can in the face of 52% uncertainty. The spirit of being data-driven involves a passion to reduce that uncertainty to 50% the next time an equivalent decision comes up.

---

*This Chapter is based, in part, on a pair of *Harvard Business Review* digital articles by Redman (2013b, 2013c).

The Real Work of Data Science: Turning Data into Information, Better Decisions, and Stronger Organizations, First Edition. Ron S. Kenett and Thomas C. Redman.
© 2019 Ron S. Kenett and Thomas C. Redman. Published 2019 by John Wiley & Sons Ltd.
Companion website: www.wiley.com/go/kenett-redman/datascience

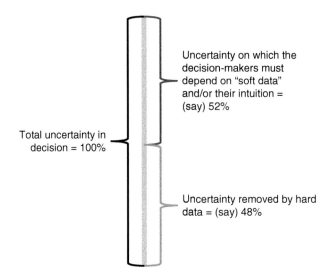

Figure 10.1   All decisions are made in the face of uncertainty. The spirit of "data-driven" involves reducing that uncertainty in the future.

This thinking is especially important for data scientists – as we've argued throughout, much of their real work involves helping people make better decisions. While advancing any single individual's and/or the organization's decision-making capability is beyond a data scientist or CAO's usual remit, it is clearly in their interest. We urge data scientists and CAOs to take it on.

We recognize that this thinking is maddeningly abstract! But over the years, we've had the good fortune to work with plenty of individual decision-makers and groups, some terrific and some simply awful. From that work, we've distilled 12 "traits of the data-driven" (Redman, 2013b) and six traits of the "anti-datas" (Redman 2013c). You can use these to baseline your organization's capabilities and identify strengths and weaknesses. In the short term, use them to help ensure your results and recommendations are listened to. And look to build greater capability in decision-makers, in the longer term.

Traits of the Data-driven

The data-driven:

- bring as much diverse data and as many diverse viewpoints to any situation as they possibly can;
- use data to develop a deeper understanding of the business context and the problem at hand;
- develop an appreciation for variation, both in the data and in the overall business;
- deal reasonably well with uncertainty, which means they recognize that they will make mistakes;
- integrate their understandings of the data and its implications with their intuitions;
- recognize the importance of high-quality data and invest in trusted sources and in making improvements;

- conduct good experiments and research to supplement existing data and address new questions;
- recognize that the criteria on which they should base a decision can vary with circumstances;
- realize that making a decision is only the first step; they know they must keep an open mind and revise decisions if new data suggests a better course of action;
- work to bring new data and new data technologies into their organization;
- learn from mistakes and help others do so; and
- strive to be a role model when it comes to data, working with leaders, peers, and subordinates to help them become data-driven.

All of these traits are important, and most are self-evident, although a few require fuller explanation. First, data-driven companies and individuals work to drive decision-making down to the lowest possible level. This may appear counterintuitive – it seems natural to seek approval at higher levels. But one executive explained the way he thought about it this way: "My goal is to make six decisions a year. Of course, that means I have to pick the six most important things to decide on and that I make sure those who report to me have the data, and the confidence, they need to make the others."

Pushing decision-making down frees up senior time for the most important decisions. And, just as importantly, a lower-level person spending hours on an issue is likely to make a better decision than an executive who spends only a few minutes. Pushing decision-making down builds organizational capability and, quite frankly, creates a work environment that is more fun to work in.

Second, the data-driven have an innate sense that variation dominates. Even the simplest process, human response, or most-controlled situation varies. While they may not use control charts, the data-driven know that they have to understand that variation if they are to truly understand what is going on. One middle manager expressed it this way: "When I took my first management job, I agonized over results every week. Some weeks we were up slightly, others down. I tried to take credit for small upturns and agonized over downturns. My boss kept telling me to stop – I was almost certainly making matters worse. It took a long time for me to learn that things bounce around. But finally I did."

Third, the data-driven place high demands on their data and data sources. They know that their decisions are no better than the data on which they are based, so they invest in quality data and cultivate sources they can trust (Redman 2016). Data scientists are well advised to earn this trust. As a result, they are prepared when a time-sensitive issue comes up. Further, high-quality data makes it easier to understand variation and reduces uncertainty. And finally, success is measured in execution, and high-quality data makes it easier for others to follow the decision-maker's logic and align to the decision.

Fourth, as decisions are executed, more data comes in. The data-driven are constantly reevaluating, refining their decisions along the way.[1] They are quicker than others to pull the plug when the evidence suggests that a decision is wrong. To be clear, it doesn't appear that the data-driven "turn on a dime"; they know that is not sustainable. Rather, they learn as they go and modify accordingly.

---

[1]Great decision-makers usually do not explicitly employ Bayes theorem, but many think like Bayesians.

Traits of the Antis

We have also distilled six bad habits that stymie managers and companies from taking full advantage of their data. We call these the traits of the "anti-datas." They include:

- preferring one's intuition over the data to an unhealthy degree;
- rigging the decision-making system (more on this in Chapter 11);
- second-guessing others;
- becoming consumed by "analysis paralysis";
- engaging in "groupthink"; and
- having deep misconceptions about data quality and/or exhibiting an unhealthy arrogance concerning the quality of one's data.

Again, most of these traits are self-explanatory, but we wish to expand on two. First, it is important to remember that the data can only take a decision-maker so far. Then, their intuitions must take over. And good decision-makers work hard to train their intuitions. At the same time, we have all met managers who say things like, "I've been working in this industry 25 years and I've seen it all. I know I can trust my gut." They are proud of their experience and are skeptical of anything new. Interestingly, we find many managers who behave this way to be solid in most respects – they care about their companies and people. They desperately want to do the right things, and they are smart. But they go to great lengths to ignore, downplay, or subvert any evidence that suggests a better way. Some even reinterpret the data to reinforce their long-held mental models. The near-certain result is decisions that are increasingly out-of-date.

Second is second-guessing. In its worst form, second-guessing involves withholding potentially useful information, then pouncing the minute a decision goes wrong. In some ways, it is natural for people competing for that next promotion to engage in second-guessing. Further, *The Forty-Eight Laws of Power* (Greene and Elffers 1998) advises those seeking power to withhold information. One observes this trait all the time in overly political individuals and companies. Data scientists should treat second-guessing as a political reality. It underscores the importance of understanding the full range of decision-making strengths, weaknesses, and biases of those you advise.

Implications

It bears mentioning that, at either the company or individual level, becoming data-driven requires deep cultural commitment, self-reflection, training, and lots of hard work. Why should a data scientist or CAO care? We have already mentioned the three most important reasons. While everyone, at both the individual and company level, likes to fashion themselves as "data-driven," the reality is far different. Decision-makers are only human, and you should help them understand their strengths and weaknesses. Start by taking a hard look in the mirror. Find someone who will tell you the truth; then rate each other on all 18 traits. You will almost certainly find it a sobering experience.

Second, we encourage you to work to advance your company's decision-making capabilities over the long term. As we've already noted, many people readily admit that "statistics was my least favorite course in college." So, start simply. We have summarized some exercises that have yielded good results for us in Chapter 12.

Finally, give decision-makers reason to trust you. Be as transparent as you can. Be open about the strengths and weaknesses in your analyses. Don't be shy when stating your recommendations. And admit it when you are wrong.

No doubt, this is hard work. But, assuming you see the value of data-driven decision-making, we ask this: "If you don't do these things, who will?" It is easy enough to bewail company politics. And incumbent on you to do something about it.

Thus, to conclude, in the near term, the real work of data scientists and CAOs involves understanding the strengths and weaknesses of decision-makers and taking them into account. In the longer term, the real work involves helping the individual decision-makers and the entire organization to become more data-driven.

11

Root Out Bias in Decision-making

Biased decision-making is the enemy of data science. We have all experienced the disappoint-ment when an important decision does not go our way.* The feeling is far worse when you feel that the decision was somehow rigged – that the decision-maker did not pay your results their due or only used some of the data. You can accept a fair decision that goes the other way, but a rigged decision feels much worse. And the ill will festers.

We have all experienced rigged decision-making in our business, civic, and personal lives. And we are not just victims. We are also perpetrators, letting bias creep into our own decisions, even if we may not realize it. Data scientists make lots of decisions – about which data to include, who to talk to, how to pose problems, which analyses to conduct, how to present results, and so forth – that impact the course of their analyses and, in turn, what decision-makers see. Even a hint of bias can have profound impact. Further, we may be complicit in ignoring, even assisting, in the biased decision-making, shading our results based on what we think decision-makers want to hear.

Rigged decisions come in many forms. Here we'll consider what to do about the most vir-ulent, which features the following steps:

1. Make the decision based on some or all of the following: intuition (see "Intuition and Rigged Decisions"), ego, ideology, experience, fear, or consul-tation with like-minded advisors.
2. Find data that justifies your decision.
3. Announce and execute the decision. Defend it to the minimum degree necessary.
4. Take credit if the decision proves beneficial or assign blame if not.

> **Intuition and Rigged Decisions**
>
> In Chapters 8 and 10, we highlighted uncer-tainty as a dominant feature of all decisions and that intuition has an important role. We further noted that some decision-makers have great intuition, but that others rely on it far too much. When this happens, the potential for rigged decisions grows.

*This Chapter is based, in part, on a *Harvard Business Review* digital article by Redman (2017b).

The Real Work of Data Science: Turning Data into Information, Better Decisions, and Stronger Organizations, First Edition. Ron S. Kenett and Thomas C. Redman.
© 2019 Ron S. Kenett and Thomas C. Redman. Published 2019 by John Wiley & Sons Ltd.
Companion website: www.wiley.com/go/kenett-redman/datascience

Data scientists must be ready to tackle this chain of events. And they need to be smart about it.

Understand Why It Occurs

Data scientists know that rigged decisions are antithetical to everything they stand for. So, approach rooting it out using data science – first try to understand it. Start with the first step: make the decision. Why do so many people make the decision first?

As we have noted, making good decisions involves hard work. Important decisions are made in the face of great uncertainty (informally, it appears to us that the more important the decision, the greater the uncertainty) and often under time pressure. The world is a complex place – people and organizations respond to any decision, working together or against one another, in ways that defy comprehension. There are way too many factors to consider. There is rarely an abundance of trustworthy data that bears directly on the matter at hand. Quite the contrary; there are plenty of partially relevant facts from disparate sources, some of which can be trusted, some not, pointing in different directions. With this backdrop, it is easy to see how one can fall into the trap of making the decision first. It is so much faster! Don't discount this benefit.

There are other reasons: Decision-makers may be motivated by how their decisions will appear to their superiors, to increase their personal power, and to pay back a favor. They may have grown overly confident in their own capabilities, or their past experiences with data and data scientists have gone poorly. There are dozens of possible considerations, and data scientists are well advised to understand the motivations of those they advise.

Once people take the first step (deciding in advance), the second step (seeking data to justify the already-made decision) comes easily enough. Decision-makers know that those impacted may ask how the decision was made, complain about it, even act to subvert it. Decision-makers know they will have to explain themselves, so getting the data needed to defend themselves is only natural.

This route is common both in business and in the world at large – so much so that Stephen Colbert coined the term *truthiness* (Wikipedia 2018c) to roughly mean the preference for concepts or facts one wishes to be true. There has always been plenty of data to support whatever decision one wants to make. And doing so has grown progressively easier with the penetration of the Internet, social media, and special interest groups. Further, it is all too easy to fall victim to confirmation bias (McGarvie and McElheran 2018), where one pays more attention to data that supports a decision and dismisses what does not.

Steps three and four (announcing the decision and either claiming credit or assigning blame) also come easily enough.

Take Control on a Personal Level

Before decrying rigged decisions made by others, we recommend that data scientists first work to improve their own decision-making. How can you avoid this trap? The first part of the answer lies in simply admitting your lack of confidence. None of us like to admit we are biased (Kahneman et al. 2011) – after all, the word carries negative connotations. But the best decision-makers we know freely admit their preconceptions. What values or beliefs may be coloring your thinking? Taking such a hard look in the mirror forces you to acknowledge other perspectives, softens your knee-jerk reaction to make a quick decision, and forces you to seek a broader view.

One preconception that too many data scientists make, particularly those early in their career, is that they must "fall in line." No matter what, they must never challenge the boss.

Although there are certainly company cultures that discourage doing so and vindictive bosses, it is the sort of preconception that you should examine (see "Bringing the Boss Bad News"). Why do you feel that way? Is there any data to support it? Are there counterexamples?

The second part advises that you reverse your inclinations: what would happen if you decided to move forward in the opposite direction from that you originally chose? Gather the data you would need to defend this opposite view and compare it to the data used to support your original decision. Reevaluate your decision in light of the more complete data set. Your perspective may still not be complete, but it will be much more balanced.

In parallel, ask yourself, "Am I the right person to decide here? Or should someone else, who has time to assemble a complete (and hence less-biased) picture, make this decision?" If so, then you should assign the decision to that person or team.

Third, before you commit to announcing, executing, and defending your choice, first try out your decision on a "friendly" or two. A friendly usually refers to someone who is on your side and wants you to succeed. Here, we are referring to someone who also wants to protect you and has the courage to tell you honestly when your thinking is incomplete, when you have missed something important, and when you are just plain wrong. If a friendly advises any of these things, then start anew, completely rethinking your decision and the data you need to make it.

Few people set out to make a rigged decision, but when you are pressured to

Bringing the Boss Bad News

Many people fear telling the boss their concerns that he or she is about to make a mistake, or they fear bringing bad news. After all, the boss may "shoot the messenger." Our experience is that it can be different. We have found that most senior managers are well aware that people do not like to do this. So they value people who will tell them the hard truths. So, of course, we are not advising that you say, "Boss, that's dumb," in front of the entire team. Instead, learn how to raise your concerns or bring bad news in a discrete and supportive manner.

make a choice quickly, you may fall victim to a flawed process. By admitting your own preconceptions, and subjecting your thinking to someone who will really challenge you, asking yourself tough questions, you can expose a rigged decision-making process. You realize just how difficult it is to completely eliminate bias. And you will make yourself a better data scientist.

Solid Scientific Footings

To develop a deeper appreciation for a scientific framework for handling bias in decision making, data scientists should study the groundbreaking work of two psychologists: Amos Tversky and Daniel Kahneman. Tversky died from leukemia in 1996 at a relatively young age, while Kahneman was awarded the 2002 Nobel Prize in Economics. They established behavioral economics, a new domain of great importance to data science. See Lewis (2017) for a popular description of the work.

A brief summary: we know now that decision-makers are affected by several mechanisms that blur their ability to properly interpret data-driven reports, including:

1. base rate neglect
2. overconfidence

3. anchoring
4. representativeness
5. availability
6. regression toward the mean
7. spurious correlation
8. framing.

   To illustrate, consider the data from two experiments (Tversky and Kahneman 1981). The value of N listed in brackets represents the number of respondents in these experiments, who were randomly assigned to Problem 1 or Problem 2.

Problem 1
(N = 152): Imagine that the United States is preparing for the outbreak of an unusual disease, which is expected to kill 600 people. Two alternative programs to combat the disease have been proposed. Assume that the exact scientific estimates of the consequences of the programs are as follows:

- If Program A is adopted, 200 people will be saved (72% made this selection).
- If Program B is adopted, the probability that all 600 people will be saved is one-third and the probability that no one will be saved is two-thirds (28% made this one).

Which of the two programs would you favor?
Now consider an alternative formulation:

Problem 2
(N = 155): Imagine that the United States is preparing for the outbreak of an unusual disease, which is expected to kill 600 people. Two alternative programs to combat the disease have been proposed. Assume that the exact scientific estimates of the consequences of the programs are as follows:

- If Program C is adopted, 400 people will die (22%).
- If Program D is adopted, the probability that no one will die is one-third, and the probability that 600 people will die is two-thirds (78%).

The expected number of deaths is the same in both problems, but people ignore this. The majority choice in Problem 2 shows a willingness to take a risk: the certain death of 400 people is less acceptable than the two-in-three chance that 600 will die. The preferences in Problems 1 and 2 illustrate a common pattern: choices involving possible benefits more often elicit risk-averse decisions, and those involving possible losses more often elicit risk-taking decisions.

   The implications for data scientists are profound – small changes in the ways you present findings can have enormous consequences. Be conscious of your own biases, and make sure they do not intrude.

   Building on this theme, Figure 11.1 is based on a famous Muller–Lyer optical illusion. On the left, the lower horizontal line seems longer. On the right, with a frame, we clearly see that the lines are equal. Just so, data scientists should make sure they frame their results fairly.

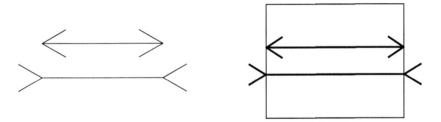

Figure 11.1   Muller–Lyer optical illusion with and without frame.

Implications

Learning how to spot and avoid bias in your own decisions will give you a new appreciation for how insidious bias is and how difficult it is to eliminate. You will not get anywhere preaching about bias while hiding your own. So, smart data scientists and CAOs strive first to lead by example.

This discussion underscores why it is so important that CAOs and, to a lesser degree, all data scientists advance a data-driven culture.

Thus, the real work of data scientists and CAOs involves lessening the bias that creeps into their own decision-making along with leading by example. Finally, as opportunity permits, the real work involves helping decision-makers recognize and then reduce their biases as well.

12

Teach, Teach, Teach

Nothing improves data science like a demanding decision-maker, one who is striving to become data-driven (Chapter 10), who wants to bring as much data and data science as possible to bear and constantly expects data scientists to deliver more. At the same time, we have already noted that many people's least favorite course in college was statistics (and they have not forgotten!). Don't be surprised when your colleagues and decision-makers are skeptical of p-values, logistic regressions, and ANOVA. Your only recourse to overcome these hurdles is to "teach, teach, teach."

You will have to work out the overall program yourself. But we can give you proven materials to get you started, including some powerful exercises; a "starter kit" of questions that will help decision-makers ask tough, penetrating questions; and, in the next chapter, a more formal template called "information quality" that they can use as they gain experience. Redman has used the material in this chapter successfully for 25 years, while Kenett and colleagues are amassing considerable success with the InfoQ template. The teaching we envision addresses the needs of individuals, teams, and organizations.

The Rope Exercise

This exercise[1] aims to show people how difficult it is to make even the simplest measurement. To complete it, you need lengths of rope, roughly 10–12 feet, one for each participant. Now take them through the following, as the pictures illustrate (Figure 12.1).

First, lay the rope out in a circle in front of you.

Second, pick up one end and cross it over so the new circle is the circumference of your waist.

Third, mark the crossover point with your hands, pick up the rope, and wrap it around your waist.

Finally, rate how well you did.

---

[1] Redman first learned of this exercise from his father, Charles Redman, in the late 1970s. He claimed to have used it often in his work at Eli Lilly, the drug company based in Indianapolis, Indiana.

The Real Work of Data Science: Turning Data into Information, Better Decisions, and Stronger Organizations,
First Edition. Ron S. Kenett and Thomas C. Redman.
© 2019 Ron S. Kenett and Thomas C. Redman. Published 2019 by John Wiley & Sons Ltd.
Companion website: www.wiley.com/go/kenett-redman/datascience

Figure 12.1   Series of steps in taking participants through the rope exercise.

Most people overestimate their waists by 50% or more. There is certain to be lots of discussion. Encourage it. And point out that there is no simpler measurement than length, and everyone knows where his or her waist is. Imagine the complications in measuring viscosity deep in a well bore, propensity to buy, or an individual's contribution to growth.

This exercise only takes 15 minutes. And most come away with a new appreciation for measurement.

The "Roll Your Own" Exercise

This exercise[2] takes participants through the full range of data science, requiring nothing more than an open mind, a pencil, a couple of sheets of paper, and a mobile phone calculator. First, advise your charges to pick something that interests, even bothers, them. Good candidates include meeting start times, caloric intake, actual time at work, and commute time. Whatever it is, participants should form it up as a question and write it down. We'll use "Meetings always seem to start late. Is that really true?" as an example throughout.

Next, ask them to think through some data that can help answer the question, and develop a plan for creating it. Have them write down all the relevant definitions and their protocol for collecting the data. For this particular example, they'll have to define when the meeting actually begins. Is it the time someone says, "OK, let's begin"? Or the time the real business of the meeting starts? Does kibitzing count?

Now they should collect the data. It is critical that they trust the data. And, as they go, they will almost certainly find gaps. They may find that although a meeting has started, it starts anew when a more senior person joins in. Advise them to modify their definitions and protocols as they go.

Next, have them draw some pictures. As we have discussed, good pictures make it easier to both understand the data and communicate main points to others. There are plenty of good tools to help, but get them to draw their first few pictures by hand. Tom's go-to plot is a time-

---

[2] Adapted from Redman (2013e).

series plot, where the horizontal axis has the date and time and the vertical axis has the variable of interest. Thus, a point on the graph in Figure 12.2 is the date and time of a meeting versus the number of minutes late.[3]

Now participants should return to the question they started with and develop summary statistics. In this case, "Over a two-week period, 10% of the meetings I attended started on time. And on average, they started 12 minutes late."

But urge your charges to go further, asking "So what?" In this case, "If those two weeks are typical, I waste an hour a day. And that costs the company x dollars a year."

Many analyses end because there is no "So, what?" Certainly, if 80% of meetings start within a few minutes of their scheduled start times, the answer to the original question is, "No, meetings start pretty much on time," and there is no need to go further.

But this case demands more. So, urge people to get a feel for variation. Note on the graph that 8–20 minutes late is typical. A few meetings start right on time, others nearly a full 30 minutes late. People are tempted to conclude that they can arrive at meetings 10 minutes late, just in time for them to start. But the variation is too great.

Now get them to ask, "What else does the data reveal?" In the example, note that six meetings began exactly on time, while every other meeting began at least seven minutes late. Odd? But upon bringing meeting notes to bear, one learns that all six on-time meetings were called by the vice president of finance. Evidently, this person starts all meetings on time!

Note that this exercise gets people to think about data, analyses, and decision-making in new ways. Urge people to push whatever example they choose to work as far as they can, always asking, "Where do I go from here?" and "Are there important next steps?" This example illustrates a common dichotomy. On a personal level, results pass both the "interesting" and "important" test. Most people would love to get back an hour a day. And while they may not be able to make all meetings start on time, they can certainly take a cue from the finance VP and start the meetings they control promptly.

On the company level, results so far pass only the "interesting test." You don't know whether these results are typical, nor whether others can be as hard-nosed as the VP when it comes to starting meetings. But a deeper look is surely in order: Are these results consistent with others' experiences in the company? Are some days worse than others? Which start later, conference calls or face-to-face meetings? Is there a relationship between meeting start time and most senior attendee? Return to step one, pose the next group of questions, and repeat the process. Keep the focus narrow – two or three questions at most.

Note that this simple exercise helps participants to experience the full range of data science work we described in Chapter 1: defining a problem, collecting relevant data, drawing some plots and exploring what the data reveal, answering the original question, posing new questions, gathering new data, and exploring implications. Most people have fun with this exercise, and many find they enjoy teasing insights from data. It also helps them gain insights into the life of a data scientist.

---

[3] The dean of the State University of New York at Binghamton School of Management, Tom Kelley, collected similar data in the context of a very successful Six Sigma project aimed at improving the functioning of the dean's office. For this data and its analysis, see Kenett et al. (2014).

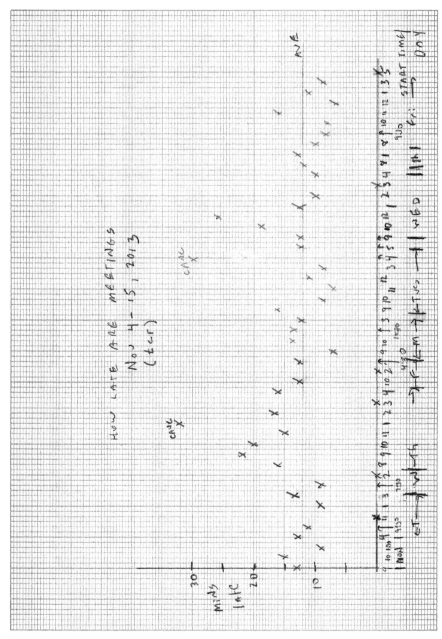

Figure 12.2 Tom's original plot of his meeting start time data.

The Starter Kit of Questions to Ask Data Scientists

Quite naturally, decision-makers do not fully trust an analysis or the results or its full implications when their understanding of data science is weak. Many learn by asking tough, penetrating questions but, for data science, most simply do not know where to start. You can help them by providing them this eight-question "starter kit" (Redman and Sweeney 2013a). These questions will also help you be better prepared!

1. *What problem are you trying to solve?* Does it align with my own? It is far too easy for data scientists (and others for that matter) to go on extended "fishing expeditions," seeking "interesting insights" that are not tethered to the business. While a certain amount of exploration is healthy, most innovation is of the small-scale, one improvement at a time variety – even with data. Encourage your data scientists to focus initially on known issues and opportunities as well as more tangible insights. As your confidence in them (or at least a few individuals) grows, give them freer rein. At the same time, you should develop a keen eye for the difference between "exploring a difficult path" and "wallowing around."
2. *Do you have a deep understanding of what the data really means?* We discussed the nuance and subtleties in data quality in Chapter 6. Unfortunately, too often people gather data without a complete understanding of the wider context in which the data was created, and misunderstandings find ways to hide themselves until it is too late. All data, even well-known quantities like "force," are subtle and nuanced. NASA (which truly has "rocket scientists") crashed a Mars lander because one team used the English measurement "foot-pounds" and another used the Metric measurement "newtons" (Pollack 1999). The potential for such problems only grows with the less familiar the data – especially social media, the IoT, automatic measurement devices, etc. – and as more intermediaries touch the data.
3. *Should we trust the data?* As also discussed in Chapter 6, untrustworthy, inaccurate data is the norm. Just as a car can be no better than its parts, so, too, analyses can be no better than the data. Some data is inherently inaccurate (GDP forecasts); other data becomes inaccurate through processing errors (Barrett 2003). All too often, data collection is just not up to snuff. For example, far too many credit reports contain inaccuracies (Bernard 2011). Unless there is a solid quality program in place, expect the data to be poor! Demand that data scientists explain how they've identified and dealt with the issues and are fully transparent about whether the data used in analyses really is "good enough."
4. *How did the analytic work go?* Some analyses proceed quickly and easily – there are a minimum of integration issues; it is obvious what the few best analytic techniques are, and they yield similar results; good graphics seem to suggest themselves; and further uses of the results come easily to mind. Other times, everything about the work is an enormous chore – the data scientist had to make too many choices about the data resolution, integration took longer than expected, and so forth. Demand that data scientists be fully transparent about their work, their level of confidence, and their intuitions about implications beyond the stated goal.
5. *Are there "big factors," preconceived notions, hidden assumptions, or conflicting data that could compromise your analyses?* There is much going on here. First, it's natural to expect a return from our investment in data and analytics, but there's a sneaky side effect. People will "find" what they think you want. Saying upfront that you expect a 10% uptick in revenue can cause people to find a short-term 10% growth that's not there in the long term or to be so busy looking for the 10% that they miss a potential 100% gain.

   Second, advanced analytics involves considerable judgment. Data scientists may have included some data sets in their analyses and excluded others. This affects the structure of the data used in the analysis and the quality of the generated information. You need to make sure they've not done so in unfair ways. The clarity and completeness of the answer correlates with the weight you should give to their conclusions.

Third, data science is essentially about developing a deeper understanding of how the world works. False assumptions are crippling. For example, the assumption that home prices were uncorrelated across markets was a major contributor to the financial crisis that began in 2007 (Silver 2012). You should insist that data scientists state their assumptions in your language – don't look away sheepishly if one states, "We've assumed 'homoskedasticity.'"

6. *Will your conclusions stand up to the scrutiny of our markets, moderately changing conditions, and a "worst-case scenario"?* Don't confuse data science with classical physics. Verifying conclusions is not as simple as repeatedly dropping uneven weights from a tower. You want your data scientists to be skeptical; to challenge each other; to test, test, and test again; and to quantify, or at least fully describe, the uncertainty in their conclusions under normal situations and to make clear when uncertainty explodes! This is critical because operationalization may well be beyond the purview of the data scientist.

7. *Who will be impacted and how?* For example, privacy is a very touchy subject – both inside and outside of your organization. The line between helpful and creepy is gray and thin (Duhigg 2012), different societies think about privacy very differently, individuals vary and change their mind, and legal frameworks (e.g. GDPR; see Appendix D) are only now under development. Data scientists can produce startling insights, but they are not fully equipped to think through the implications. It's important to ask this question not only to the data scientists you work with but your colleagues, privacy specialists, and those charged with protecting the company's brand as well.

8. *What can I do to help?* Quite obviously, there is no need to ask this question if the answers to the first seven questions don't satisfy. Bear in mind here that any important discovery will have implications across the organization. We're particularly concerned about change management. All change is difficult, and resistance to counterintuitive results will prove far too strong for most data scientists.

HBR Guide to Data Analytics Basics for Managers

We are understandably reluctant to tell anyone to "go read something," although we make one exception for this book. A compilation of digital articles by luminaries such as Tom Davenport, D. J. Patil, and Michael Schrage; journalists such as Amy Gallo; and a host of others, it is simply outstanding. The chapters are short, focused, and extremely well written. Managers should scan it and put it on their shelves for reference as the need arises. And data scientists should study it to understand where decision-makers are coming from.

The starter kit, as the name implies, is broad and not particularly deep, although it will almost certainly lead to deep discussions. It facilitates discussions on the range of topics that both data scientists and decision-makers consider. At the same time, decision-makers should bring in other experts on topics such as privacy as the situation demands.

Implications

We've previously opined (Chapter 10) that helping individual decision-makers and the entire organization become increasingly data-driven is in the data scientists' and CAOs' best long-term interests. The waist measurement with rope aims to help decision-makers appreciate how difficult even the simplest measurement can be and is usually great fun. The second exercise aims to help participants develop a deeper appreciation for data science and is considerably more demanding. Finally, the starter set of questions aims to help decision-makers ask good questions. In the short term, this experience may not be so pleasant. But the long-term benefits, as decision-makers become more comfortable, both with specific analyses and data science more generally, may prove enormous.

In parallel, they can also improve everyone's quantitative sophistication. Think about this for a moment – imagine how powerful a company would be if everyone, across the entire company, did just a little more basic data science every day! You will probably benefit from a comprehensive program covering variation and the distinction between correlation and cause and effect, A/B testing, and contemporaneous topics specific to HR, marketing, drilling for oil, and so forth. We'll build on these exercises in the next chapter, which focuses on the InfoQ framework.

To conclude, the real work of data scientists and CAOs involves helping colleagues and decision-makers grow more comfortable with data science. Thus, teach them the basics in easy, engaging ways when you can and confront more difficult issues as opportunity permits.

13

Evaluating Data Science Outputs More Formally

In the last chapter, we focused on teaching your colleagues some basics and providing a starter set of questions for decision-makers. Of course, this business of helping decision-makers become increasingly better consumers of data science never ends. As they gain experience, you need to provide them a more formal template based on the eight dimensions of the information quality model (Kenett and Shmueli 2016a). This will help them go deeper, facilitate discussions regarding trade-offs, and help them improve the quality of information generated in their organizations. Breiman (2001) depicts two cultures in the use of statistical modeling to reach conclusions from data, data modeling, and algorithmic analysis. The InfoQ framework addresses outputs from both approaches, in the context of business, academic, services, and industrial work.

Assessing Information Quality

The InfoQ framework provides a structured approach for evaluating the analytic work. InfoQ is defined as the utility, U, derived by conducting a certain analysis, f, on a given data set, X, with respect to a given goal, g. For the mathematically inclined:

$$\mathrm{InfoQ}(U,f,X,g) = U\big(f(X|g)\big).$$

As an example, consider cellular operators who want to reduce churn by launching a customer retention campaign. Their goal, g, is to correctly identify customers with high potential for churn – the logical target of the campaign. The data, X, consists of customer usage, lists of customers who've changed operators, traffic patterns, and problems reported to the call center. The data scientist plans to use a decision tree, f, which will help him or her define business rules that identify groups of customers with similar churn probabilities. The utility, U, is increased profits by targeting this campaign only on customers with a high churn potential.

The Real Work of Data Science: Turning Data into Information, Better Decisions, and Stronger Organizations,
First Edition. Ron S. Kenett and Thomas C. Redman.
© 2019 Ron S. Kenett and Thomas C. Redman. Published 2019 by John Wiley & Sons Ltd.
Companion website: www.wiley.com/go/kenett-redman/datascience

InfoQ is determined by eight dimensions that can be assessed individually in the context of the specific problem and goal. These dimensions include the following:

1. *Data resolution.* Are the measurement scale, measurement uncertainty, and level of data aggregation appropriate relative to the goal?
2. *Data structure.* Are the available data sources (including both structured and unstructured data) comprehensive with respect to the goal?
3. *Data integration.* Are the possibly disparate data sources properly integrated together? Note: this step may involve resolving poor and confusing data definitions, different units of measure, and varying time stamps.
4. *Temporal relevance.* Is the time frame in which the data was collected relevant to the goal?
5. *Generalizability.* Are results relevant in a wider context? In particular, is the inference from the sample population to the target population appropriate (statistically generalizable – Chapter 8)? Can other considerations be used to generalize the findings?
6. *Chronology of data and goal.* Are the analyses and needs of the decision-maker synched up in time?
7. *Operationalization.* Are results presented in terms that can drive action?
8. *Communication.* Are results presented to decision-makers at the right time and in the right way (as described in Chapter 7)?

See Appendix C for a detailed list of questions used in InfoQ assessments.

Importantly, InfoQ helps structure discussions about trade-offs, strengths, and weaknesses. Consider the cellular operator noted above and consider a second potential data set X*. X* includes everything X has, plus data on credit-card churn, but that additional data won't be available for two months. Resolution (the first dimension) goes up, while temporal resolution (the fourth) goes down. Or suppose a new machine learning analysis, f*, has been conducted in parallel, but results from f and f* don't quite line up. "What to do?" These are the most important discussions for decision-makers, data scientists, and CAOs.

Further, the InfoQ framework can be used in a variety of settings, not just for helping decision-makers become more sophisticated. It can also be used to assist in the design of a data science project, as a midproject assessment, and as a postmortem to sort out lessons learned. See Kenett and Shmueli (2016a) for a comprehensive discussion of InfoQ and its applications in risk management, health care, customer surveys, education, and official statistics.

A Hands-On Information Quality Workshop

This workshop uses InfoQ to help an entire team understand the importance of clear goals and what it takes to achieve information quality with respect to those goals. It combines individual work, team discussions, and group presentations, using the InfoQ framework.

Phase I: Individual Work
Please complete the following four steps and document the results of each for further discussion.

Step 1: The Background
Pick an organization to focus on. It should be one that you know reasonably well, such as your current or previous place of employment, a school, hospital, or restaurant.

1. Answer the following: Who are this organization's most important customers and suppliers? What are its most important products and services?
2. Select an important goal for this organization. It can be reducing costs, improving quality, or gaining new customers. This step defines the g and U components in the InfoQ equation.

Step 2: The Data
List various data sources that are available to help decision-makers pursue that goal. In evaluating data sources, focus on data quality and data clarity. Data quality reflects the extent to which the data can be trusted, and data clarity represents the way data elements are defined and collected by various parts of the organization. This step specifies the X component of InfoQ.

Step 3: The Analysis
Identify several approaches for analyzing the data in order to help the organization achieve its goal. In this step, identify and list alternative methods of analysis, $f_1, f_2, ..., f_p$.

Step 4: Assessment
Assess the data and the potential analysis on eight InfoQ "dimensions" using a 1–5 score where 1 means "very poorly" and 5 "very well."

1. *Data resolution.* When the data is on the right level of granularity, the measurement scale is appropriate, and the level of aggregation is appropriate, score a "5."
2. *Data structure.* When there are important gaps in the data coverage, score a "1."
3. *Data integration.* A "5" corresponds to integration into a seamless whole.
4. *Temporal relevance.* When the data is timely with respect to the goal, score a "5."
5. *Generalizability.* When what we learn can be generalized to many other circumstances, score a "5."
6. *Chronology of data and goal.* When the analysis and recommendations can be completed in a timely fashion from a decision-making perspective, score a "5."
7. *Operationalization.* If the analyses are unlikely to lead to concrete actions that provide business benefit, score a "1."
8. *Communication.* If the "who" (needs the information), "what," "when," "why," and "how" are clear, score a "5."

Note: an application for recording InfoQ scores, which also allows for a range of values reflecting uncertainty in the score, is available for download from the Wiley website of Kenett and Shmueli (2016a). The application requires installation of the JMP software and provides an overall InfoQ score based on the geometric mean of the individual dimension scores.

Phase II: Teamwork
Form teams of three or four participants.

1. Share your case study with the team and engage in open discussion.
2. Choose the case study that your team will present in the group session.
3. Prepare your case study.

Phase III: Group Presentation
Each team delivers a presentation on its selected case study.

1. Workshop participants rate the presentation using the eight dimensions presented earlier.
2. Summarize all ratings and engage in open group discussion.

Implications

This chapter aims to help you continue the process of teaching decision-makers that we started in the previous chapter. Decision-makers evaluate data scientists and their work every day, even if only informally. Best to help them understand more formal criteria (e.g. InfoQ) on which to do so. Finally, even the best data science involves trade-offs, and formal criteria can help you hold the right discussions more proactively.

We can't say it frequently enough – the real work of data scientists and CAOs involves helping decision-makers become better, more demanding consumers of data science.

14

Educating Senior Leaders

Imagine senior executives with no particular interest in data and data science. Yet, every day, they receive a cacophony of grandiose proclamations that "data is the new oil," extravagant claims about analytics and artificial intelligence, exhortations to become data-driven, heated claims that they must digitize, dire warnings about privacy and the reputational damage that results from data breaches. Yes, about five tech firms are profiting, but no important competitor has truly embraced data and data science. The cover of the May 6, 2017, issue of the *Economist* proclaimed that data is now "the world's most valuable asset," but these claims have been around a long time. Further, the failure rate on analytics projects is high (Demurkian and Dai 2014). The messages conflict and none of them fit together.

Closer to home, the senior executive team still can't get sales figures from three systems to agree after two years of trying. Complicating matters still further, the chief information security, chief privacy, chief information technology, chief data, chief digital, and chief analytics officers and others have narrow perspectives and compete for attention and resources. They are certainly not lying, but no one can present a clear picture. No straight answers anywhere.

Senior executives in this situation could only conclude that the data space is a confused and overhyped mess. And they are correct!! With a trusted perspective (or data) missing, senior executives are likely to revert to their intuition.

Clearly this does not bode well for data in general or data science in particular. At best, it means that data science won't necessarily get asked to contribute on the really important problems and will be underfunded, will probably be ill-placed, and will not be given a fair chance. It means that any sort of data or analytics transformation must wait for the next regime, as transformations must be led from the top. At worst, it imperils the entire company. The list of "once great companies" is littered with those who didn't spot and act on an industry-changing technology, idea, or capability quickly enough.

Of course, most senior leaders are not so dismissive, and many know they have to sort it all out. But one should expect confusion, gaps, and some misconceptions. CAOs should make it their personal missions to clear up the confusion, fill the gaps, correct misconceptions, and provide perspective. Senior executives, including board members, are entitled to simple, complete, unbiased explanations. It bears mention that CAOs are uniquely qualified to provide these explanations – after all, the job involves seeing the essential simplicity in complexity.

The Real Work of Data Science: Turning Data into Information, Better Decisions, and Stronger Organizations, First Edition. Ron S. Kenett and Thomas C. Redman.
© 2019 Ron S. Kenett and Thomas C. Redman. Published 2019 by John Wiley & Sons Ltd.
Companion website: www.wiley.com/go/kenett-redman/datascience

Thousands of questions can come up. At the very least, CAOs should be prepared to address the following:

- What all is involved?
- Where does analytics (and everything else for that matter) fit?
- What are the must dos for senior executives?

Finally, CAOs must develop the trust to be asked these questions and the perspective to answer them directly.

Covering the Waterfront

Interestingly, oil, in and of itself, does very little for anyone. You can use it to power engines, but even then, fuel and an engine get you little. Much more is required. You have to pull a number of components together into a product (say, a car); you have to design and manufacture the car; you have to sell it; and you need a company to accomplish all these things together.

We find that the car, and a car company, provide a useful framework for extending the "data is the new oil" analogy. See Figure 14.1.

Let us consider several features of the analogy in turn. First, security is to data as safety is to the car. Thus, manufacturers work hard to make the car safe, with airbags, bumpers, crumple zones, power-assisted brakes, and blind-spot mirrors. Likewise, companies must employ a wide variety of techniques to keep data *secure*.

Design and manufacture

Governance: Well-managed manufacturing

Digitization: Latest manufacturing technologies

Metadata: Inventory management

Marketing and sales

Monetization: Sold, serviced at profit

Running the company

Infonomics: Improving the balance sheet

Data as an asset: Managing facilities and intellectual property

Figure 14.1   Data in the context of a car: what it takes to manufacture and sell it and to run a successful automobile company.

Similarly, technology is the new engine. The engine powers the car and, without technological advances, a data- and analytics-led transformation would not be possible. Technologies include databases, communications equipment and protocols, applications that support the storage and processing of data, and the raw computing horsepower, much of it now in the "cloud," to drive it all. It is odd that this infrastructure is referred to as the "cloud." After all, much of it consists of fiberoptic cable, deep underwater.

In the analogy, we liken the data-driven concept to the car's GPS. As discussed in Chapter 10, "data-driven" means bringing as much data as you can to support decisions. GPS does just that for drivers, helping them navigate to their destinations using the latest traffic and accident reports.

Continuing, the *Internet of Things* is a catchall term for embedded and connected devices, built into products and manufacturing lines, that create new data and effect control. The Nest home thermometer is an example, and in the analogy we liken them to devices built into cars that make them easier to maintain. Eventually, one can implement conditioned-based maintenance (CBM), where the maintenance schedule reflects the car condition and driving patterns, not a one-size-fits-all approach. CBM can both reduce cost and improve safety. It requires, however, sensor data and analytic models with predictive capabilities.

Similarly, we liken privacy – the notion that people have rights to control the use of facts about themselves – to tinted glass, which can keep the identities of those inside the car from others. Frankly, this analogy is weak. Privacy is complex, as different people feel very differently about their privacy. The situation is rapidly changing, and the European GDPR portends further change all over the world (see Appendix D).

It bears mention that cars are increasingly online, integrated, and autonomous. So the security, data, control, and privacy concerns in the car mirror those of any other company. For example, security is not just airbags and bumpers – the real fear is cybersecurity – someone hacking into and hijacking the car.

Other important concepts in the data space parts lead to analogies involving design and manufacture, marketing and sales, and running the company. First, manufacturing. In the data space, "digitization" or "digital transformation" refers to the application of digital technologies, wherever possible, to operations and decision-making. Digitization builds on existing technologies with the goal of increasing scale and decreasing unit cost. We liken it to the use of advanced robotics in automobile manufacturing. Chapter 17 is about the industrial evolution toward advanced manufacturing.

Just as manufacturing is well controlled, the concept behind data governance is that all aspects of data operations, including moving it around, its usage, creating it, and changing it, should be well controlled.

Next up is metadata, which we have already mentioned. Metadata is data that assists in the interpretation of other data. Examples include data definitions and data models. The topic can be incredibly esoteric. We mention metadata here because so many issues, such as the inability to provide consistent answers to basic questions from three systems, stem from inadequate metadata and, in turn, from imprecise business language. We liken it to the inventory management system in the factory.

One topic that we feel to be underappreciated is proprietary data. Unlike most employees, who can be hired away, or capital – your dollar is the same as everyone else's – your data is uniquely your own. If you manage well, some of it may gain "proprietary status" and will be a great source of competitive advantage. Well-known examples include Facebook's *friend,* LinkedIn's *connection,* and Standard and Poor's *CUSIP.* In our analogy, we liken proprietary data to intellectual property the auto manufacturer uses to develop a unique product.

Next come marketing and sales. Data monetization is the notion that data can be "sold" or "licensed" at profit or built into other products that can be sold at profit. It is directly analogous to automotive marketing and sales. The notion that data should be treated as an asset and managed as professionally and aggressively as other assets (Redman 2008) has gained considerable traction in the past decade. For a car company, this means they manage data on par with capital. Infonomics takes the idea a step further, positing that data should appear on the balance sheet (Laney 2017).

To complete the analogies, we note that, in the picture above, we likened data to fuel and high-quality data to high-quality fuel. But of course "data" appears over and over in other parts of the analogy.

So, too data science, which includes descriptive, predictive, and prescriptive analytics; visualization; statistics; AI; machine learning; natural language process; and business intelligence. Properly deployed, data science makes everything better!

The term *big data* is also used, and misused, frequently. Properly, big data involves volumes, varieties, and velocities of data that cannot be processed by traditional means. Although we do not believe a car, on its own, qualifies, managing an entire fleet may well entail managing big data.

The car analogy is especially useful in what is called "directed imagination." This well-known approach has been used in several contexts, most impressively in the training of athletes. For example, ice skater Elizabeth Manley and diver Greg Louganis claim that imagery helped them win their Olympic medals. Data scientists and CAOs certainly need to be imaginative, and the car analogy and directed imagination tools can help. This topic is beyond the scope of our book, however.

Companies Need a Data and Data Science Strategy

As the car analogy makes clear, there is much going on in the data space. Beyond the confusion it engenders, the hype makes things appear easier than they really are in the data space. Hire a chief data officer and a few data scientists, put your data in the cloud, turn algorithms loose, and reap the benefits in reasonably short order.

This is simply wrong! There is no holy grail, instant pudding, or quick win with data. With the possible exception of data quality (more below), everything in the data space is hard work. Companies have much to learn in coming up to speed and should plan for a certain amount of trial and error.

In summary, there are lots of ways to profit from data and much that can go wrong. It leads us to conclude that every company needs a data strategy, fully integrated with its overall business strategy. The strategy should embrace the company's current and desired industry position, competitive landscape, how and where the company wishes to compete with data, proprietary data, personnel, and tolerance for risk. Companies must make hard choices, so there are certainly areas where the right approach is "wait and see." But those choices should be made on the basis of solid work, not inattention!

Quite frankly the toughest issue is talent. Companies can recover if they are a bit late in adopting a new technology. But talent, especially data scientists and the ability to manage them, are in short supply and will be for the foreseeable future.

Organizations Are "Unfit for Data"

Today's organizations are "unfit for data" (Redman 2013d). As used here, "organization" consists of four components, along with an example or two of companies' lack of fitness for data:

- *People*. Companies lack enough people with the needed skill and expertise up and down the organization chart.
- *Structure*. Silos are the enemy of data sharing, impeding the cross-functional coordination and work.
- *Policy and control*. Roles and responsibilities for data are misaligned.
- *Culture*. Even though they say they do, people and organizations do not value data and data science.

There is a great deal here, so we will only pursue two points. First, we find that many people confuse data and technology. In the past, this led companies to assign responsibility for data to their information technology departments, to the detriment of both. But data and technology are very different kinds of assets and require different styles of management.

Importantly, technology is increasingly a commodity. As Nicholas Carr (2003) pointed out more than 15 years ago, basic storage, processing, and communications technologies are readily available to all, at a fraction of the cost of just a few years ago. If anything, the trends Carr called out are accelerating – witness the stunning progress of cloud computing, the penetration of mobile, and easy access to advanced analytic and AI techniques.

These points lead us to conclude that the first step is to separate management responsibilities for data and technology.

The second point is finding the right spots for data scientists, and too many companies, perhaps unwittingly, set their data scientists up to fail (Redman 2018a). We take this topic up in the next chapter.

Get Started with Data Quality

We took up data quality in the context of data science in Chapter 6. Across a company, most data is in poor shape (Nagle et al. 2017), and the associated costs are enormous (think 20% of revenue; Redman 2017c). Worse, people rightly do not trust the data (*Harvard Business Review* 2013), and you certainly cannot expect them to make a data-driven decision if they do not trust the data. We find that most data issues have rather simple roots and can be eliminated with relative ease. Thus, data quality is a great place to start a data program. And the savings will provide the funds for everything contemplated here.

Implications

Educating senior management and helping guide overall data strategy is a tall order indeed. The space is a confused mess and the topic is very charged and political. There are always good reasons to delay, or simply avoid, the tough issues. But if a company or agency wishes to enjoy more of the benefits data and data science offer, CAOs have no choice. Thus, the real work of CAOs is building trust and gravitas so they will be listened to, sorting through the many perspectives on data, leading the discussions necessary to help senior managers understand the real issues, and helping chart a course.

15

Putting Data Science, and Data Scientists, in the Right Spots

Maintenance is a big area for analytics. One data scientist, employed at an international semiconductor company, develops optimal "split maintenance schedules." The idea is to perform maintenance tasks in several steps, separated in time, replacing major maintenance shutdowns. The advantage is less total downtime, which translates into huge savings in wafer fabrication. He enjoyed great success within the maintenance organization and hoped to build upon that success by combining it with CBM, which employs production systems and raw material data. But this data is only available from the operations department's databases. And he couldn't get access. As the story illustrates, "Silos are the enemy of data sharing." They are particularly damaging to data science because so many opportunities lie in combining data.

The Need for Senior Leadership

Organization Structure Affects Data Science

This chapter pays homage to W. Edwards Deming (1900–1993). Deming, a physicist turned statistician turned management consultant, had incredible impact, first in Japan and later in the West. Many in Japan credit Deming for the Japanese postwar economic miracle of 1950–1960, and his impact on the rest of the world is incalculable. At the core, Deming believed that improving quality and productivity requires a fundamental transformation, based on widespread deployment of statistical thinking. And he advised that this transformation also required an organizational component. Today's data science is no less transformational.

Deming called for a very senior "leader in statistical methodologies," which herein we call the chief analytics officer.

The only solution, in today's hierarchical, command-and-control organizations, lies in a CAO who sits high enough in the management chain and has the authority and personal gravitas to insist on data sharing. W. Edwards Deming (see "Organization Structure Affects Data Science") called for such a solution in the early 1980s: "There will be a leader of statistical methodology, responsible to top management. He must be a man of unquestionable ability.

The Real Work of Data Science: Turning Data into Information, Better Decisions, and Stronger Organizations,
First Edition. Ron S. Kenett and Thomas C. Redman.
© 2019 Ron S. Kenett and Thomas C. Redman. Published 2019 by John Wiley & Sons Ltd.
Companion website: www.wiley.com/go/kenett-redman/datascience

He will assume leadership in statistical methodology throughout the company. He will have authority from top management to be a participant in any activity that in his judgment is worth his pursuit. He will be a regular participant in any major meeting of the president and staff" (Deming 1986). While today's CAO must rely on personal gravitas more than formal authority, Deming was surely on the right track.

At a minimum, data science should be aligned with a company's most important strategic priorities. For example, Carlo Torniai, head of data science and analytics at Pirelli, the tire manufacturer, concentrates on three main areas: smart manufacturing, cybertechnologies, and the extended value chain, from the supply of raw materials to the final point of sale. His team's brief includes measuring and managing the data more precisely and using real-time information to develop more efficient solutions in these domains (Pirelli 2016).

For Pirelli, the biggest source of data is the production line. It measures the operational parameters associated with tire making and the quality of the product throughout. For any tire, Pirelli monitors the raw materials used as inputs and the different settings and readings on the machines that produced it. Armed with this information, Pirelli builds predictive models for the expected quality of that tire.

The next step is to move from predictive to prescriptive models and adjust machine settings in real time. The system will "learn" each time it makes a change and, as a result, the process will be continually improved.

Pirelli's data team aims to add technologies that predict when tire maintenance is needed, letting drivers know when their tires should be inflated, repaired, or changed. This allows fleet managers to keep downtime to a minimum.

"In the not-so-distant future we envision a virtual factory where at any given time the allocation of resources and expected outcome is known, and where machines can automatically regulate processes and material flow and suggest skills required on the floor," says Torniai (Pirelli 2016).

Torniai feels that part of his work is to explain this new approach while proving the business case. "It doesn't just require technical skills but communication skills and the ability to tell stories with data to people who are not necessarily technical folks," he says. "Then it's about explaining that often you don't get a black and white solution but a range of possibilities. So you need to explain the 'fuzziness' in results to people who are used to dealing with straight numbers, then use this as the basis to make decisions."

Torniai could not be successful without the full confidence of senior management.

Building a Network of Data Scientists

Further, as we have previously noted, the need for data science in every sector, in every company, and in every department therein is becoming increasingly clear. Companies, smart ones anyway, are just beginning to realize that a fundamental transformation, driven by data, increased computing power, and burgeoning AI capabilities, is afoot. The most important part of the CAO's job is ensuring that the company has a network to well-placed data scientists to support, and in some cases lead, this transformation.

The worst mistake a company can make is to hire a cadre of smart data scientists, perhaps organized into a data science lab, provide them with access to the data, and turn them loose, expecting them to come up with something brilliant (Redman 2018a). Lacking focus and support, most fail.

Data Science Centers of Excellence: Can Data Science Be Outsourced?

Many companies group their data scientists together in "centers of excellence." Reasons for doing so include providing a better environment to help data scientists grow into their roles, cultivate their craft, and learn from one another; to resolve funding issues; and to create a critical mass of talent. Data scientists assume roles much like internal consultants (although, of course, they are not just advising but doing the work). This is a good option for some companies and problems.

Extending this logic, the center of excellence could just as easily be a separate company that decision-makers hire on a fee-for-service basis. And given the (relative) newness of data science, it is only natural for senior leaders to ask if they should do so and outsource their data science efforts. After all, learning how to manage data science is a tall order. Further, a cottage industry of companies offering data science services is emerging. Such companies already know how to manage the effort and already have a cadre of seasoned data scientists and plenty of experience on hand. We see some merit in the approach, particularly for companies that are just getting started, and the problem at hand demands expertise they simply do not have. But we do not think it is a good long-term solution. Whether in-house or not, companies still have to manage the effort. More importantly, data science is increasingly becoming a source of competitive advantage. Sooner or later, companies must learn how to grow and retain the talent. It is never a good idea to outsource your competitive advantage.

In contrast, Deming proposed putting statisticians in the line: "There will be in each division a statistician whose job it is to find problems in that division, and work on them. He has the right and obligation to ask questions about any activity of the division, and he is entitled to responsible answers" (Deming 1982).

While we concur with Deming's thinking, it needs to be adapted to current needs and technologies. We actually see a continuum of data science opportunities. On one end are opportunities for basic process improvements. For such problems, putting data scientists "in the line" is clearly appropriate and in accord with Deming (see also Hahn 2007).

On the other end of the spectrum are more speculative opportunities (e.g. rethinking credit decisions based on social media data) that require fundamental innovation. These must be performed in a "data lab" (Redman and Sweeney 2013a).

And of course, as Figure 15.1 suggests, there are opportunities that occupy a middle ground in this continuum. Each requires its own structure. For example, fine-tuning a sophisticated algorithm might be best done in a "center of excellence," which some companies adopt (see "Data Science Centers of Excellence: Can Data Science Be Outsourced?").

There is no "one-size-fits-all" solution. Rather, CAOs must balance the needs for data science to be close to the decision-makers they support, the company's ability to manage data scientists (few line managers really know how to treat data scientists), and practical political realities.

Further, CAOs are unlikely to have full control over all the data scientists in even a relatively small company – as managers are free to hire their own data science teams (plenty of data scientists report into marketing, for example). Thus, the watchword for CAOs is ensuring that decision-makers are supported, through a mix of embedded, outsourced, and laboratory-based data scientists, while building a community among them.

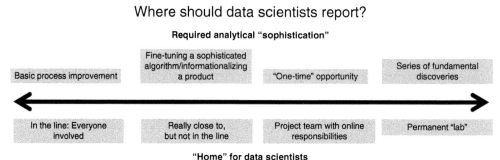

Figure 15.1 The best "home" for data scientists, organizationally, depends on the sort of problem they are attempting to solve.

Implications

The real work of the CAO involves acquiring enough data science talent and getting the right organizational structure in place so the overall team of data scientists (and the organization as a whole) is most effective in turning data into information, making better decisions, and building organizational capabilities. The CAO must have the trust of senior executives if this is to happen, so building that trust, and cultivating personal gravitas, is his or her most important job.

The CAO should report as high into the company as possible. He or she will also lead a contingent of data scientists, variously placed according to their roles. Some, such as those helping solve day-in, day-out problems, will reside in the line. Conversely, a few devoted to longer-term or more strategic work may reside in a "data lab." Finally, those whose work supports midterm objectives may reside in a "center of excellence" or other structure close to, but not in, the line.

Finally, the real work of the CAO involves adjusting the organizational structure more or less continuously, as everything in the data space is in continuous flux.

16

Moving Up the Analytics Maturity Ladder

Let us say you work for a bank, car manufacturer, or cellular operator. Your company completes many tasks and transactions designed to meet the needs of its customers and accumulates written reports, call center transcripts and recordings, engineering specs, financial data, inventory levels, and many other types of data. What you and the company do with this data reflects the maturity level of your company's management approach and analytic capabilities.

Understanding this maturity level helps you do a better job leading data science efforts, both short and long term. We distinguish between five maturity levels:

Level 1. Firefighting: random reports to be delivered yesterday.
Level 2. Inspection: a focus on descriptive statistics.
Level 3. Process view: modeling variability with statistical distributions.
Level 4. Quality by design: planning interventions and experiments for data gathering.
Level 5. Learning and discovery: a holistic view of data science.

Here we expand on the characteristics of these levels and their impact on data science. Going up the maturity ladder provides deeper and wider benefits from data and data science. This parallels the effect known as the "statistical efficiency conjecture."[1]

Let's start with firefighting. *Firefighting* reflects a heroic level of maturity of organizations, so chaotic that people can't think beyond the short term. Most firefights don't require much data analysis, as fires are visible and generate flames, heat, and smells. Firefighters need to figure out the problem immediately and provide quick-fix solutions. The work is frantic! Most companies cannot stay in firefighting mode long. Products and services don't measure up, and both customers and employees leave. The data scientist, in such environments, is asked for reports to be produced yesterday. These are typically shallow and lack insight. They are

---

[1] The statistical efficiency conjecture states that as organizations move up these levels, they become more efficient at solving problems, offering better products and services at lower costs. This conjecture has been tested with 21 case studies (Kenett et al. 2008).

The Real Work of Data Science: Turning Data into Information, Better Decisions, and Stronger Organizations,
First Edition. Ron S. Kenett and Thomas C. Redman.
© 2019 Ron S. Kenett and Thomas C. Redman. Published 2019 by John Wiley & Sons Ltd.
Companion website: www.wiley.com/go/kenett-redman/datascience

shallow, because they focus only on immediately available data, and they lack insight, because management does not give data scientists an opportunity to reflect on their findings.

Inspection provides a way out of firefighting. To prevent problems from reaching the customer, companies inspect every product and activity. Inspection data – for example, from the IoT, such as usage tracking and in-line process control applications – helps determine the quality of a product or service. When collected over time, inspection data provides a rearview mirror perspective.

Organizations at the inspection maturity level have much data to analyze. Business intelligence platforms, such as Power BI, Tableau, and the pivoting capabilities in Excel, allow data scientists to visualize this data in various perspectives and slices. The typical report in such organizations is based on dashboards with descriptive statistics such as bar charts and pie charts.

Just as it is impossible to drive a car by looking in the mirror, so too it is difficult to run an organization with historical data only. Drivers need to look ahead, through the windshield, and organizations require a similar look-forward capability. The science of good prediction took a big leap forward with the control chart, invented by Walter Shewhart at Bell Laboratories in 1924. Shewhart explicitly embraced variation in his formulation. Critically, the chart triggers an alarm when a process is out of control, provides a platform for improvement, and often helps identify those opportunities.

Imagine you are in 1924 and you work for a company developing, producing, installing, and maintaining the telephone system in America. Your boss asks you to provide to the factory floor a tool for managing the telephone assembly line. The idea is that instead of relying on mass inspection, you are asked to develop a tool that helps control the production process. In this context, Walter Shewhart wrote to his boss on May 16, 1924: "The attached form of report is designed to indicate whether or not the observed variations in the percent of defective apparatus of a given type are significant; that is, to indicate whether or not the product is satisfactory."

Shewhart did not stop there: "The theory underlying the method of determining the significance of the variations in the value of p is somewhat involved when considered in such a form as to cover practically all types of problems." The control chart proposed by Shewhart extends to many other domains (Shewhart 1926; Kenett et al. 2014). It is a great example of the contributions data scientists and statisticians can make. The control chart addresses a real problem, provides a practical tool for operators and managers, and is theoretically sound. Figure 16.1 is an example from a modern web-based system.

In this spirit, in 2016, at a major semiconductor company, data scientists integrated data from the wafer production line with testing data to determine how much additional testing was required. Thus, chips whose supporting data indicate they are more likely to be defective are tested in greater depth, and those deemed less likely to be defective undergo lesser scrutiny. It's a "win-win-win," saving time, improving quality, and contributing to the bottom line. This reflects 90 years of advances in industrial statistics.

Organizations with such a *process view* require data collection, data analysis, and data presentation that include predictive analytics and online monitoring. This requires a data scientist with significantly greater capabilities. The control chart is based on the concept of a statistical distribution representing the performance of a stable process. It indicates when the underlying distribution has changed and the process has gone "out of control." The data scientist, in such an environment, works with probability distributions in the background.

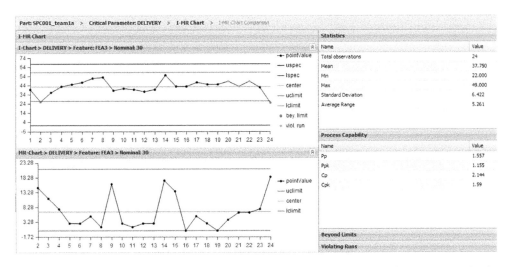

Figure 16.1 An example of a control chart tracking individual measurements (upper chart tracks the measurements, lower chart their moving range representing variability). The right panel provides summary statistics. Source: Figure from SPCLive system by KPA Ltd.

Ambitious companies move up to the next level of the data analytics maturity curve, *quality by design,* when managers extend this forward-looking thinking into the design of products and services. This stage requires statistically designed experiments, robust design, and other methods to ensure the product/service meets customer needs and performs well, even when the raw materials are of uneven quality and environmental conditions vary. Genichi Taguchi and Joseph M. Juran played key roles in developing this approach. In the 1980s, Taguchi introduced the West to methods he developed in Japan in the 1950s (Taguchi 1987). Juran described a structured approach for quality planning that started with understanding customer needs and ensuring those needs were met in the final product (Juran 1988).

In quality by design organizations, the data scientist understands the role of experimental design and is becoming proactive in the planning of interventions. This is a precursor of A/B testing (Kohavi and Thomke 2017), where web application designers direct customers to alternative designs to understand which works best, using data such as click-through rates.

In service companies, which offer human-intensive services, there is natural predisposition to perform quality by design in a more qualitative fashion, because of the following:

- A/B testing is not allowed for compliance/regulatory reasons, or there are policies requiring that all customers must be treated equally.
- Customer relationships are an important aspect of the service, which is provided by combining several products and/or services.
- Customers are so different from each other that A/B testing is too difficult.

For more on behavioral big data, see Shmueli (2017).

Neither Taguchi nor Juran anticipated the big data era, with data coming from all quarters, including social media, web clicks, "connected devices" (e.g. the IoT), personal trackers, and so forth. This new age poses both new challenges and opportunities and suggests to us a fifth maturity level, which we call *learning and discovery*. The opportunities are enormous, including personalized medicine, optimized maintenance, data-driven decision-making, and so forth.

Very savvy data scientists are needed. For example, a data scientist may harvest data from social media and integrate it with operational reports to produce deep insights and establish causal relationships. But is the social media data biased because of self-selection? What roles can A/B tests play? What is the nature of statistical inference when the data sets are so large? And, most critically, how does data science drive business strategy? Data scientists need to answer such questions.

Figure 16.2 presents the five maturity levels with a brief description of how data is used at each level.

In proposing this maturity ladder, we emphasize the need for organizations to move up to the learning and discovery level. Senge (1990) has emphasized the importance of doing so. This maturity ladder parallels the quality ladder, matching management style with industrial statistics methods as proposed by Kenett and Zacks (2014). Information quality dimensions, important at the fifth level of maturity, were described in Chapter 13. Organizations at the learning and discovery maturity level are good at generating information of high quality.

Two further remarks. First, one should not expect all parts of an organization to be at the same level of maturity. Some individuals and departments will lead, others will lag. Further, even in the best-run companies, crises do occur. So, there is no "one-size-fits-all" approach to data science.

Second, "data," as an asset in and of itself, is asserting itself more and more (Redman 2008). After all, consumers need data, as do knowledge workers, decision-makers, and data scientists. And, as noted in Chapter 6, much of this data is in poor shape (Nagle et al. 2017). CAOs should see both risk and opportunity (Kenett and Raanan 2011).

Level 5: *Learning and discovery* - This is where attention is paid to information quality. Data from different sources is integrated. Chronology of data, goal and generalization is a serious consideration in designing analytic platforms.

Level 4: *Quality by design* - Experimental thinking is introduced. The data scientist suggests experiments, like A/B testing, to help determine which website is better.

Level 3: *The process focus* - Probability distributions are part of the game. The idea that changes are statistically significant, or not, is introduced. Some attention is given to model fitting.

Level 2: *The descriptive statistics level* - Management asks to see histograms, bar charts, and averages. Models are not used, data is analyzed in rather basic ways.

Level 1: *Random demand for reports driven by firefighting* - New reports address questions such as: how many components of type X did we replace last month or how many people in region Y applied for a loan?

Figure 16.2   The analytics maturity ladder.

Implications

So why does all this matter? Three reasons. First, the real near-term work for CAOs involves recognizing where the company/division you serve falls on the maturity curve and building a team suited to work at that level. For example, you will not have done your job if you build a team with great depth in AI when the company is struggling to establish basic control. Your data scientists will grow frustrated and leave for greener pastures, and the company will continue to struggle. There is a hard lesson here – everyone wants to work on the latest and greatest. But you have to match the focus of your team to the organization's maturity, not the other way around.

Second, the midterm and the work there is even tougher. You have to move your company up a level to derive greater insights from numbers (Kenett 2008, 2017). This is no easy task and you may fail. But take heart and ask yourself who is better qualified to lead such an effort than you. Indeed, we would argue that data scientists are uniquely qualified in this regard. For example, early in this chapter, we opined that there is no organized data to analyze when the company is in firefighting mode. This is not quite true – the simple observation about how many fires there are to fight can be revealing. Of course, you will need to fully integrate yourself into the business, including all of its politics, to find this out. So what? Everything important is political. Get on with it.

Finally, there is data quality. Most companies, even level five manufacturers, are at the firefighting or inspection levels when it comes to data quality. Those who've gotten to the process level enjoy far better data at far lower costs. Look here for opportunity. Ultimately, the goal is to reach the learning and discovery maturity level, where data science and data scientists reach their full potential.

Thus, the real work of CAOs involves establishing a team suited for the organization's current level of maturity in the short term and leading efforts to move up the maturity ladder in the longer term.

17

The Industrial Revolutions and Data Science

This chapter provides an industrial context into the advance of data science. This context is important because it illustrates an important role for CAOs and senior managers – namely, identifying macrotrends based on evidence from disparate sources and positioning their companies to take advantage of them.

Consider five work environments:

1. *Craft activities*. An artisanal system where workers learn their trade through apprenticeship. Learning accumulates through experience, and little data is kept.
2. *Repetitive activity*. Machinery makes it possible to complete repetitive work more efficiently. The iconic breaokthrough is the water-powered loom, and it characterizes the first industrial revolution. Data from inspections is used to certify products.
3. *The factory*. Fully contained factories utilizing repeatable processes produce better, cheaper goods for the masses. These characterize the second industrial revolution. Data is used for process control.
4. *The automated factory*. In the third industrial revolution, computers started managing processes. In principle, a suite of integrated applications supports the entire factory, from managing inventory to tracking work orders.
5. *Industry 4.0*. The fourth industrial revolution is unfolding now. It features unprecedented levels of data, including sensors that track heat expansion, vibration, and noise in production, cyberphysical systems, IoT devices, and advanced analytics to process it all.

The term *Industry 4.0* is applied in many contexts, for example, Healthcare 4.0, Hotel Industry 4.0, Food 4.0, and Education 4.0. All follow trajectories that parallel those taken in manufacturing.

The Real Work of Data Science: Turning Data into Information, Better Decisions, and Stronger Organizations, First Edition. Ron S. Kenett and Thomas C. Redman.
© 2019 Ron S. Kenett and Thomas C. Redman. Published 2019 by John Wiley & Sons Ltd.
Companion website: www.wiley.com/go/kenett-redman/datascience

The First Industrial Revolution: From Craft to Repetitive Activity

In medieval Europe, most families and social groups made their own goods such as cloth, utensils, and other household items. The only saleable cloth was woven by peasants who paid their taxes in kind to their feudal lords. Barons affixed their marks to the fabric, which came to stand for their levels of quality. While some details differ, the textile industry all over Europe and China was similar. It was apparently the first industry to analyze data. Simple production figures, including percentages of defective products, were compiled in British cotton mills early in the nineteenth century. Quality control activities generated data that was aggregated in ledgers for accounting and planning purposes (Juran 1988).

The industrial revolution started in England. Richard Arkwright (1732–1792) was an English inventor and a leading entrepreneur who became known as the "father of the modern industrial factory system." He invented the spinning frame and a rotary carding engine that transformed raw cotton into cotton lap. Arkwright's achievement was to combine power, machinery, semiskilled labor, and a new raw material, cotton, to create mass-produced yarn. In 10 years, he became the richest man in England.

The Second Industrial Revolution: The Advent of the Factory

During the early twentieth century, a constellation of technologies and management techniques expanded mass production. The internal combustion engine (and the oil and gas needed to fuel it) and electricity powered the way, the production line formalized the division of labor, and huge factories were built. The Taylor system, featuring time and motion studies, drove production tasks and productivity quotas. And companies learned how to manage enormous factories (Chandler 1993). This was the second industrial revolution.

As one example, Western Electrics' Hawthorne Works, on the outskirts of Chicago, employed up to 45,000 workers and produced unheard of quantities of telephone equipment and a wide variety of consumer products. It was in this environment that Shewhart realized that manufacturing processes can be controlled using control charts (Shewhart 1926). Control charts minimized the need for inspection, saving time and money, and delivering higher quality. W. Edwards Deming and Joseph M. Juran were instrumental in bringing this approach to Japan in the 1950s. Deming emphasized the use of statistical methods (Deming 1931), and Juran developed a comprehensive management system featuring the so-called quality trilogy (Godfrey and Kenett 2007). Like Shewhart, both worked for Western Electric in the late 1920s.

From a data analysis perspective, attention shifted from inspection to process and the need to understand variation. Thus, statistical models and probability played a key role.

The Third Industrial Revolution: Enter the Computer

Computers have changed manufacturing in several ways. We've picked three to illustrate.

First, computers enabled "mass customization" (Davis 1997). Essentially, mass customization combines the scale of large, continuous-flow production systems with the flexibility of a job shop. This allows a massive effort, with batches of size one. A call center that employs screening to route calls to the right specialists is a good example.

Second is automation of so-called back-office functions, such as inventory management and product design. Take the development of an automobile suspension system designed using computer-aided design. The new suspension must meet customer and testing requirements under a range of specific road conditions. After coming up with an initial design concept, design engineers use computer simulation to show the damping effects of the new suspension design under various road conditions. The design is then iteratively improved based on these results.

Third is integration. Thus, in parallel to the design of the suspension system, purchasing specialists and industrial engineers proceed with specifying and ordering the necessary raw materials, setting up the manufacturing processes, and scheduling production using computer-aided manufacturing (CAM) tools. Then, throughout manufacturing, tests provide the necessary production controls. Finally, CAM pulls everything together. Ultimately, of course, the objective is to minimize the costly impact of failures in a product after delivery to the customer. Computer simulations required new experimental designs, including Latin Hypercubes and Kriging models. In addition, modern advances in optimization of statistically designed experiments have led to new designs that better address constraints and exploit optimality properties (for details on these methods, see Kenett and Zacks 2014).

The Fourth Industrial Revolution: The Industry 4.0 Transformation

We are now in the midst of the fourth industrial revolution, fueled by data from sensors and IoT devices and powered by increasing computer power. Information technology, telecommunications, and manufacturing are merging, and production is increasingly autonomous. Futurists talk of machines that organize themselves, delivery chains that automatically assemble themselves, and applications that feed customer orders directly into production.

There are many implications for data scientists. According to IDC (2018), the top analytical technologies include:

- natural language generation, natural language processing, and text mining
- speech recognition
- virtual agents
- machine learning platforms
- AI-optimized hardware
- decision management.

We emphasize three common Industry 4.0 themes of special relevance to data scientists: data quality (Chapter 6), information quality (InfoQ (Chapter 13), and the need to move quickly up the analytics maturity ladder (Chapter 16).

Implications

It is trite to observe that we are in a period of rapid technological change. If anything, we expect the pace of change to accelerate. Look how much more quickly the fourth industrial revolution followed the third, and the third followed the second! Further, change stems not just from technology but governments, wars, and people. The rise of China and the advance of gay rights over the past generations are examples. Those who do not keep up risk losing out.

Thus, the real work of CAOs involves having long-term perspective, a historical context, and a view to the future. This means identifying long-term trends based on evidence from disparate sources and weaving them into a compelling narrative. This also means helping position their companies to be ready. For CAOs employed by companies that make high-tech products, there is a special consideration. For years, companies have outsourced manufacturing, concentrating on design. This may be dangerous as Industry 4.0 advances, because continued outsourcing may threaten the company's advantages in design. CAOs should study this issue carefully.

18

Epilogue

"Much fine work in statistics involves minimal mathematics; some bad work in statistics gets by because of its apparent mathematical content" (Cox 1981, p. 295). This down-to-earth description of work done in statistics, by one of the most important statisticians of this century, is quite telling. Sir David Cox started his career as a statistician at the Wool Industries Research Association in Leeds, England. The experiences he gained in this work environment shaped his extensive groundbreaking work in statistical methodology.

Our key message is that the real work of data science focuses on solving important problems facing people, companies, and organizations while fully embracing the complexities, preconceptions, bad data, quirks of decision-makers, and politics that go with them. This real work must be rooted in sound theory, or the solutions will not hold up for long. This combination touches on various domains, some methodological, some technological, some organizational, and some personal. The 18 chapters in this book cover the nontechnical ones. To paraphrase Cox's quote: the real work of data science requires a holistic approach, beyond computational algorithms and machine learning technologies.

Another famous statistician, John W. Tukey raised the flag half a century ago, calling for a serious discussion on the future of statistics (Tukey 1962). That future has arrived, and it is called "data science."

Strong Foundations

Statistics researchers have pursued the reasoned understanding of data sets for decades. Among the core discoveries were sampling methods and sufficiency properties, which provide data scientists the technical foundations for dealing with very large data sets. Research on topics as diverse as generalized regression and shrinkage estimators continues to put new tools in the data scientist's quiver.

Technical advances produced significant game changers. Appendix E provides a brief overview of recent technical advances. These methods leverage the ever-growing availability of big data, featuring many variables and different data structures. These advances have also raised new ethical concerns, which we discuss in Appendix D.

The Real Work of Data Science: Turning Data into Information, Better Decisions, and Stronger Organizations,
First Edition. Ron S. Kenett and Thomas C. Redman.
© 2019 Ron S. Kenett and Thomas C. Redman. Published 2019 by John Wiley & Sons Ltd.
Companion website: www.wiley.com/go/kenett-redman/datascience

With respect to our topic, a breakthrough came when statisticians realized that data analysis involves a range of concerns, beyond mathematical properties of statistical tools. To quote John Tukey:

> For a long time I have thought I was a statistician, interested in inferences from the particular to the general. But as I have watched mathematical statistics evolve, I have had cause to wonder and to doubt....All in all I have come to feel that my central interest is in data analysis, which I take to include, among other things: procedures for analyzing data, techniques for interpreting the results of such procedures, ways of planning the gathering of data to make its analysis easier, more precise or more accurate, and all the machinery and results of (mathematical) statistics which apply to analyzing data.

An amazing insight, now over 50 years old. Continuing, Tukey notes that:

> data analysis is a very difficult field. It must adapt itself to what people can and need to do with data. In the sense that biology is more complex than physics, and the behavioral sciences are more complex than either, it is likely that the general problems of data analysis are more complex than all three. It is too much to ask for close and effective guidance for data analysis from any highly formalized structure, either now or in the near future. Data analysis can gain much from formal statistics, but only if the connection is kept adequately loose. (Tukey 1962)

Going back still further, as early as the 1930s, W. Edwards Deming wrote in the preface of Shewhart's book on *The Economic Control of Quality of Manufactured Product*:

> Tests of variables that affect a process are useful only if they predict what will happen if this or that variable is increased or decreased. Statistical theory, as taught in the books, is valid and leads to operationally verifiable tests and criteria for an enumerative study. Not so with an analytic problem, as the conditions of the experiment will not be duplicated in the next trial. Unfortunately, most problems in industry are analytic. (Deming 1931)

The most important parts of data science are analytic (e.g. predictive), addressing the concern voiced by Deming.

These are the foundations on which we build in this book.

A Bridge to the Future

Our goal is to help data scientists navigate the individual and organizational complexities. Experienced data scientists will recognize the points made in these 18 chapters. We hope not to have scared newcomers off – but be aware that your technical knowhow is just table stakes.

The book explores, in short chapters, the things data scientists aren't usually taught in class, but that the giants of statistics knew were essential. Of course, solid analyses are essential, but it is other, more complex steps that data scientists must take to ensure their analyses are given their due, lead to good decisions, and produce results. This stuff is messy, and we've introduced several models to help simplify. We proposed a life-cycle model and the organizational ecosystem right up front, in Chapter 1, and explored each step in subsequent chapters. For example, we urge data scientists to invest in understanding the business they serve and the real problems they must confront, attack data quality proactively, bring both hard and soft data to bear, and present their results in simple, powerful ways. We introduced practical statistical efficiency as a means for data scientists to assess and improve their impact.

We took a hard look at what it means to be data-driven (Chapter 10) and explored dealing with bias in decision-making (Chapter 11). We urged data scientists to raise everyone's game by teaching their colleagues and decision-makers to become more educated, demanding customers of data science (Chapters 12 and 13) and CAOs to help senior leaders understand the complexities of the data landscape (Chapter 14). We urged CAOs to find the right organizational homes for data science (Chapter 15), and we explored the analytic maturity ladder (Chapter 16) as a means to mark progress and improve still further.

All this to help today's data scientist become more effective today.

We're even more interested in the not-too-distant future. Data and data science can be a force for transformational good in all aspects of human endeavor, from making us all freer and safer, to promoting equality, to better health care at lower cost, and to economic growth and shared prosperity. But it would be naïve, even reckless, to assume this will happen on its own. Data and data science are completely agnostic – data doesn't care whether it is simply wrong, and algorithms don't give a whit whether they invade privacy or promote social injustice. It is time for data scientists and CAOs to step up to their real work.

Appendix A

Skills of a Data Scientist

This list builds on Joiner (1985) and Kenett and Thyregod (2006), mapping skills to the life cycle introduced in Chapter 1.

General:

- Have a genuine desire to solve real problems and help others make sound decisions.
- Learn new computing environments and applications quickly.
- Be a good problem solver.
- Commit to, and meet, deadlines.
- Recognize and deal with your own personal biases.
- Recognize ethical issues and deal with them effectively (see Appendix D).
- When needed, have the courage to represent an unpopular point of view in an appropriate way.

1. Problem elicitation:

- Help others formulate their problems and opportunities.
- Listen carefully and ask probing questions.
- Distinguish the important problems from those of lesser significance.

2. Goal formulation:

- Learn the problem domain and speak the domain language.
- Make good estimates of how much effort will be required to solve the problem.
- Meet decision-makers regularly on their home ground.
- Network effectively.

3. Data collection:
- Participate in, or at least observe, data collection.

The Real Work of Data Science: Turning Data into Information, Better Decisions, and Stronger Organizations,
First Edition. Ron S. Kenett and Thomas C. Redman.
© 2019 Ron S. Kenett and Thomas C. Redman. Published 2019 by John Wiley & Sons Ltd.
Companion website: www.wiley.com/go/kenett-redman/datascience

4. Data analysis:

  - Have a broad knowledge and true understanding of scientific methods.
  - Adapt existing analytic procedures to novel environments.
  - Keep abreast of developments in data science.
  - Use the simplest appropriate analytic procedure to get the job done.

5. Formulation of findings:

  - Explain results in the language of decision-makers.
  - Work with decision-makers to ensure they understand the important nuances and uncertainties associated with results.

6. Operationalization of findings:

  - Support implementation of important decisions in practice.

7. Communication of findings:

  - Teach in ways that best suit those you are teaching.
  - Learn to convince others of the validity of a solid solution and see to it that proper action is taken.
  - Be diplomatic and know when to bend, when to stand firm, and how to help resolve conflicts among other team members.
  - Communicate effectively both orally and in writing.

8. Impact assessment:

  - Judge your work on the basis of its impact.
  - Learn from successes and failures.

Appendix B

Data Defined

There are many approaches to defining *data*. Here we use the one that best corresponds to the way data is created and used in organizations and, in our view, best supports data science (Redman 2008). Thus, *data* consists of a *data model* and a *data value*.

Data models are abstractions of the real world that contextually define what the data are all about, including specifications of things of interest (called "entities"), important properties of those things (fields or attributes), and relationships between them. Thus you, the reader, are an entity, and your employer is interested in you as an EMPLOYEE, with attributes such as DEPARTMENT, SALARY, and MANAGER. REPORTS TO is an example of a relationship. You are not just an EMPLOYEE, but also a TAXPAYER, a PATIENT, and a USER, created by the tax agency, your medical provider, and tech companies with their own interests in mind.

As mentioned in Chapter 1, data are not just numbers – data exists in the context of a data model and the purposes of the organization that defined that model. The model exists in other contexts as well, including who created it and for what reason. And putting data in its proper context with respect to any given analysis is critical for data scientists.

Today, there is much interest in *unstructured data*, which we prefer to think of as *data that has not yet been structured*.

We use the term *metadata* to refer to data that makes other data easier to use. *Data models, data definitions*, and *business rules* (that constrain data values) all qualify. It bears mention that data that has been computerized is often more useful, though the definition carries no such requirement. We also use the term *soft data* to refer to sights, smells, sounds, impressions, feelings, conversations, unstructured data, and the like that are not necessarily *hard data* but are relevant to the analysis or decision at hand. When smells are measured with electronic noses, they become *hard data*. When social media such as tweets is subjected to sentiment analysis using text analytics, like smells, it moves from soft data to hard data.

There are also many approaches to defining *information*. We find it most powerful to define information not in terms of what it *is* but in terms of what it *does*. To illustrate, suppose you are playing a game of chance with one die. You bet a dollar and select a number, 1–6. A "dealer" then rolls a die, and you either lose your bet or are paid six. You don't get to see

The Real Work of Data Science: Turning Data into Information, Better Decisions, and Stronger Organizations,
First Edition. Ron S. Kenett and Thomas C. Redman.
© 2019 Ron S. Kenett and Thomas C. Redman. Published 2019 by John Wiley & Sons Ltd.
Companion website: www.wiley.com/go/kenett-redman/datascience

the roll. Your chances of winning are roughly 1 in 6, assuming the game is fair. Now consider the following "information" about the next roll:

- *Scenario a*: Someone tells you the die is loaded and the next roll will come up odd. You'll pick 1, 3, or 5, and your chances of winning increase to 1 in 3. You've been informed.
- *Scenario b*: Someone tells you the die is loaded and will come up odd when it will really come up even. You've been misinformed.
- *Scenario c*: Someone tells you that the dealer is spinning a roulette wheel, not rolling a die. Your chances of winning are greatly reduced, but your understanding of the game comes closer to reality. You will almost certainly try to withdraw your bet. You've been informed.
- *Scenario d*: Someone tells you that the die is red. Nothing changes. You've been neither informed nor misinformed.

Information, then, teaches you about the world. Sometimes it does so by reducing your uncertainty about future events, other times by enlarging your perspective. Defining information based on the reduction of uncertainty, such as occurs in this test scenario, has a rich tradition. Claude Shannon (1948) of Bell Labs first introduced the notion for communications and developed a measure for the quantity of information, based on how much the uncertainty was reduced. Bayesian statisticians also use this concept.

Two subtleties are frequently important. First, although information can indeed be derived from data, it can arise in other ways as well. A train whistle that warns you of an approaching train is certainly informative. It qualifies as soft data, but it is hardly data (yet anyway). Second, information is intensely personal. For example, the person standing next to you, having seen the approaching train, views the whistle as an annoying blast, not information.

Appendix C

Questions to Help Evaluate the Outputs of Data Science

Kenett and Shmueli (2014) defined information quality (InfoQ) as the utility derived from a specific analysis of a specific data set, conditioned on the analysis goals. InfoQ is determined by eight dimensions discussed in Chapter 13. Questions to help assess these dimensions for a specific report are listed below (see also Kenett and Shmueli 2016a).

Dimension	Questions
1. Data resolution	1.1 Is the data scale aligned with the stated goal?
	1.2 How reliable and precise are the measuring devices or data sources?
	1.3 Is the data analysis suitable for the data aggregation level?
2. Data structure	2.1 Is the type of data used aligned with the stated goal?
	2.2 Are data integrity details (corrupted/missing values) described and handled appropriately?
	2.3 Are the analysis methods suitable for the data structure?
3. Data integration	3.1 Is the data from multiple sources properly integrated? If so, what is the credibility of each source?
	3.2 How is the integration done? Are there linkage issues that lead to crucial information being dropped?
	3.3 Does the data integration add value in terms of the stated goal?
	3.4 Does the data integration cause any privacy or confidentiality concerns?
4. Temporal relevance	4.1 Are data collection, data analysis, and deployment time-sensitive?
	4.2 Does the time gap between data collection and analysis cause any concern?
	4.3 Is the time gap between the data collection and analysis and the intended use of the model (e.g. in terms of policy recommendations) of any concern?

The Real Work of Data Science: Turning Data into Information, Better Decisions, and Stronger Organizations, First Edition. Ron S. Kenett and Thomas C. Redman.
© 2019 Ron S. Kenett and Thomas C. Redman. Published 2019 by John Wiley & Sons Ltd.
Companion website: www.wiley.com/go/kenett-redman/datascience

Dimension	Questions
5. Generalizability	5.1 Is the stated goal statistical or scientific generalizability?
	5.2 For statistical generalizability in the case of inference, is there a clear answer to the question "What population does the sample represent?"
	5.3 For generalizability in the case of a stated predictive goal (predicting the values of new observations; forecasting future values), are the results generalizable to the to-be-predicted data?
	5.4 Does the report provide sufficient detail for the type of needed reproducibility and/or repeatability, and/or replicability?
6. Chronology of data and goal	6.1 If the stated goal is predictive, are all the predictor variables expected to be available at the time of prediction?
	6.2 If the stated goal is causal, do the causal variables precede the effects?
	6.3 In a causal study, are there issues of endogeneity (reverse causation)?
7. Operationalization	7.1 Are the measured variables themselves of interest to the study goal, or is the focus on their underlying construct?
	7.2 What are the justifications for the choice of variables?
	7.3 Who can be affected (positively or negatively) by the findings?
	7.4 What can the affected parties do about it?
8. Communication	8.1 Are the descriptions of the goal, data, and analysis clear?
	8.2 Is the level of description appropriate for the decision-maker?
	8.3 Are there any confusing details or statements that might lead to confusion or misunderstanding?

Appendix D

Ethical Considerations and Today's Data Scientist

Privacy and ethical considerations have always been of concern for data scientists, but the increasing breadth and depth of available data raises these concerns to new levels. This is especially true for behavioral big data (BBD) that captures human and social actions and interactions in increasing breadth and level of detail. Such concerns include:

1. Data collection can change the subjects' behaviors.
2. The study design and analysis can harm subjects and put them at risk in intangible ways.
3. The study design is complicated by free will.
4. With time, data properties can change in response to evolving inquiry or examination methods.

Developing data analytic applications for autonomous cars provides an excellent example. In designing collision avoidance software that uses real-time data acquisition methods and pattern recognition algorithms, the data scientist may be faced with alternatives having ethical implications. Should the car, faced with an inevitable crash, hit the kid running fast from the right or the older person moving slowly from the left? There are no other options and both choices are terrible. These issues are not typically addressed by internal review boards that are compulsory in clinical research. For a comprehensive treatment of these issues, see Shmueli (2017).

Today, data scientists are front and center in growing concerns about privacy and data protection. "Privacy" is a slippery concept, with considerable national, generational, and individual variation in perspective. Ultimately, the idea is that individuals ought to have a fair say in how data about them is used. For example, one individual may encourage marketers to individualize the ads he or she sees based on search history but not wish to have political parties do the same thing. And the 2018 Facebook/Cambridge Analytica scandals have added an urgency and political dimension to the discussions (https://en.wikipedia.org/wiki/Facebook%E2%80%93Cambridge_Analytica_data_scandal).

The Real Work of Data Science: Turning Data into Information, Better Decisions, and Stronger Organizations, First Edition. Ron S. Kenett and Thomas C. Redman.
© 2019 Ron S. Kenett and Thomas C. Redman. Published 2019 by John Wiley & Sons Ltd.
Companion website: www.wiley.com/go/kenett-redman/datascience

New regulations, including the EU's GDPR and the United States' updated Common Rule (called the Final Rule), impact the use of data about people. Both regulations relate to protecting individuals' rights to determine how their private information is used, in turn impacting and affecting data collection, access, and data movement within the same country/region and between industries. GDPR and the Final Rule try to modernize what today constitutes "private data" and data subjects' rights and balance them against "free flow of information between countries." These regulations, in areas such as health-care management systems and social media, already have significant impact on the work of data scientists (Shmueli 2018), and these impacts will only grow. Finding out, after the fact, how to comply with such regulations is obviously not a good idea.

Political campaigns are another domain where data science can have severe repercussions. Election surveys not only provide information, they can also impact voter choices and whether voters go to the polls. In this and other contexts, the ability to target fake news is a reality data scientists need to face. For more on election surveys, see Kenett et al. (2018).

Further, as we described in Chapter 6, data scientists must always worry whether the data they analyze can be trusted. These concerns are exacerbated by the access control, data anonymization, and privacy-preserving sharing arrangements needed to keep data private (see Srivastiva et al. 2019). Thus, for data deemed private, data scientists may face an added layer of complexity when it comes to trust and quality.

So what should data scientists, CAOs, and the organizations that employ them do? At the very least, they must know and follow all relevant law. Further, we think they should do more. Nearly a generation ago, some futurist (name unknown) opined that "privacy will be to the Information Age what product safety was to the Industrial Age." And in that realm, most societies opted for greater consumer protections. Thus, we believe data scientists should strive to conduct their work in an ethical manner, even if what that means is not always clearly spelled out.

We recommend three sources. First, Singer (2018) describes courses given in leading universities to educate students in ethical considerations. Cornell University, for example, introduced a data science course where students learn to deal with ethical challenges such as biased data sets that include too few lower-income households to be representative of the general population. Students are also challenged to debate the use of algorithms to help auto-mate life-changing decisions like hiring or college admissions.

Second, a set of comprehensible ethical guidelines for statistical work was prepared by the committee on professional ethics of the American Statistical Association (ASA 2016). The guidelines identify six groups of stakeholders and list responsibilities of ethical statisticians. These guidelines are very broad and also have applicability to data science.

Finally, O'Keefe and O'Brien (2018) provide an even more comprehensive perspective that is applicable not just by data scientists but by all data professionals, everyone who touches data, and senior management. It is a great place to start.

Appendix E

Recent Technical Advances in Data Science

We have taken the position throughout this book that data scientists must do much more than technical work. Still in our view, it is critically important that data science builds on solid theoretical and technical foundations. So, while a full review is beyond scope, a few remarks on technical aspects of data science are in order.

Fisher (1922) laid the foundations for statistics as a discipline. He considered the object of statistical methods to be reducing data into the essential statistics, and he identified three problems that arise in doing so:

1. specification – choosing the right mathematical model for a population;
2. estimation – methods to calculate, from a sample, estimates of the parameters of the hypothetical population; and
3. distribution – properties of statistics derived from samples.

Since then, others, some quoted in our book, have built on these foundations. Of particular relevance here, Tukey (1962) envisioned a data-centric development of statistics. Huber (2011) and Donoho (2017) celebrated the 50th anniversary of Tukey's paper with an outlook on the role of statistics and reference to data science. Data science, which pulls together domain knowledge, computer science/IT, and statistics, builds further still, and today's data scientist has a large variety of powerful methods, such as regression, ANOVA, visualization, Bayesian methods, statistical control, neural networks (e.g. machine learning), bootstrapping and cross validation, cluster analysis, text analytics, logistic regression, structural equation models, time-series analysis, decision trees, association rules, and so on, at his or her disposal.

We are hopeful that data science will continue to grow, with better and better methods for analyzing data; extracting the essential information; interpreting, presenting, and summarizing results; and making valid inferences. These in turn will require better technical methods for accessing, manipulating, storing, searching, and sorting data.

The Real Work of Data Science: Turning Data into Information, Better Decisions, and Stronger Organizations, First Edition. Ron S. Kenett and Thomas C. Redman.
© 2019 Ron S. Kenett and Thomas C. Redman. Published 2019 by John Wiley & Sons Ltd.
Companion website: www.wiley.com/go/kenett-redman/datascience

Particularly important is the continuing development of AI, machine learning, and deep learning. Neural networks, first introduced in the 1980s, lie at the core. These highly parameterized models and algorithms, inspired by the architecture of the human brain, can often develop powerful predictive models when they are trained with enough high-quality data.

We are also encouraged by recent technical advances in statistical learning. This area is motivated by increasing abilities to make large quantities of measurements automatically, producing "wide data sets." While there may be huge quantities of data, the number of independent variables still far exceeds the number of observations. In text analytics, for example, a document is represented by counts of words in a dictionary. This leads to document-term matrices with 20,000 columns, one for each distinct word in the vocabulary. Each document is represented by a row, and a cell contains the number of times a word appears in the document. Most entries are zero.

In many studies, data scientists have hundreds of independent variables that can be used as predictors in a regression model (say). It is likely that a subset will do the job well, whereas including all independent variables will degrade the model. Hence, identifying a good subset of variables is essential. Statistical learning, advanced by Efron, Friedman, Hastie, Tibshirani, and others, has made it possible to handle such data sets. Efron and Hastie (2016) provide a beautiful account of computer-age statistical inference from past, present, and future perspectives. Statistical learning describes a wide range of computer-intensive data analytic algorithms now on the data scientist's workbench. Short descriptions of some specific methods follow.

First is the least absolute shrinkage and selection operator (LASSO). It is a regression-based method that performs both variable selection and regularization to enhance the prediction accuracy and interpretability of the statistical model. Other approaches based on decision trees are random forests and boosting. Decision trees create a model that predicts the value of a target variable based on several input variables. Trees are "learned" by splitting the data into subsets based on an input variable. This is repeated on each derived subset in a recursive manner called "recursive partitioning." The recursion is complete when the subset at a node has all the same value of the target variable or when splitting no longer adds value to the predictions. In random forests, one grows many decision trees to randomized versions of the training data and averages them. In boosting, one repeatedly grows trees and builds up an additive model consisting of a sum of trees. For more, see Hastie et al. (2009) and James et al. (2013). Further, for a contemporaneous discussion of current trends in data science trends, see https://mathesia.com/home/Mathesia_Outlook_2019.pdf.

It is clear that the great data scientists we discussed in Chapter 2 have exciting futures as they continually learn and leverage new statistical and other methods.

References

Aigner, M. and Ziegler, G. (2000). *Proofs from the Book*, third edition, Springer-Verlag Berlin Heidelberg, Germany.

American Statistical Association (ASA). (2016). Ethical guidelines for statistical practice. http://www.amstat.org/ASA/Your-Career/Ethical-Guidelines-for-Statistical-Practice.aspx.

Baggaley, K. (2017). Your memories are less accurate than you think. https://www.popsci.com/accurate-memories-from-eyewitnesses.

Bapna, R., Jank, W., and Shmueli, G. (2008). Consumer surplus in online auctions. *Information Systems Research* 19: 400–416.

Barkai, J. (2018). Predictive maintenance: myths, promises, and reality. http://joebarkai.com/predictive-maintenance-myths-promises-and-reality.

Barrett, L. (2003). Hospital revives its dead patients. *Baseline*. http://www.baselinemag.com/c/a/Projects-Networks-and-Storage/Hospital-Revives-Its-QTEDeadQTE-Patients. February 10.

BBC Scotland. (2017). Voice recognition lift. https://www.youtube.com/watch?v=J3lYLphzAnw.

Bernard, T.S. (2011). Are serious errors lurking in your credit report? *New York Times*. https://bucks.blogs.nytimes.com/2011/06/07/are-serious-errors-lurking-in-your-credit-report. June 7.

Box, G.E.P. (1980). Beer and statistics Monday night seminar (reported from memory by first author).

Box, G.E.P. (1997). Scientific method: the generation of knowledge and quality. *Quality Progress* 30: 47–50.

Box, G.E.P. (2001). An interview with the International Journal of Forecasting. *International Journal of Forecasting* 17: 1–9.

Breiman, L. (2001). Statistical modeling: the two cultures. *Statistical Science* 16(3): 199–231.

Camoes, J. (2017). 12 ideas to become a competent data visualization thinker. https://excelcharts.com/12-ideas-become-competent-data-visualization-thinker.

Carr, N. (2003). IT doesn't matter. *Harvard Business Review*, May, pp. 41–49.

Chandler, A. (1993). *The Visible Hand: The Managerial Revolution in American Business*. Belknap Press.

Cobb, G.W. and Moore, D.S. (1997). Mathematics, statistics, and teaching. *American Mathematical Monthly* 104: 801–823.

Coleman, S. and Kenett, R.S. (2017). The information quality framework for evaluating data science programs. In: *Encyclopedia with Semantic Computing and Robotic Intelligence* (ed. P. Sheu), 125–138. World Scientific Press.

The Real Work of Data Science: Turning Data into Information, Better Decisions, and Stronger Organizations, First Edition. Ron S. Kenett and Thomas C. Redman.
© 2019 Ron S. Kenett and Thomas C. Redman. Published 2019 by John Wiley & Sons Ltd.
Companion website: www.wiley.com/go/kenett-redman/datascience

Cox, D.R. (1981). Theory and general principle in statistics. *Journal of the Royal Statistical Society, Series A* 144(2): 289–297.

Data Science Association. (2018). Code of conduct. http://www.datascienceassn.org/code-of-conduct.html.

Davenport, R.H. and Patil, D.J. (2012). Data scientist: the sexiest job of the 21st century. Harvard Business Review. https://hbr.org/2012/10/data-scientist-the-sexiest-job-of-the-21st-century.

Davis, S. (1997). *Future Perfect*. Basic Books.

De Veaux, R., Agarwal, M., Averett, M. et al. (2017). Curriculum guidelines for undergraduate programs in data science. *Annual Review of Statistics and Its Applications* 4: 15–30. https://www.annualreviews.org/doi/abs/10.1146/annurev-statistics-060116-053930.

Deming, W.E. (1931). Dedication to The Economic Control of Quality of Manufactured Product by W. Shewhart, pp. i–iii. D. Van Nostrand Company, Inc.

Deming, W.E. (1982). *Quality, Productivity and the Competitive Position*. Cambridge, MA: MIT, Center for Advanced Engineering Study.

Deming, W.E. (1986). *Out of the Crisis*. Cambridge, MA: MIT Press.

Demurkian, H. and Dai, B. (2014). Why so many analytics' projects still fail. *Analytics*. http://analytics-magazine.org/the-data-economy-why-do-so-many-analytics-projects-fail. July/August.

Donoho, D. (2017). 50 years of data science. *Journal of Computational and Graphical Statistics* 26(4): 745–766.

Doumont, J.L. (2013). Creating effective slides: design, construction, and use in science. https://www.youtube.com/watch?v=meBXuTIPJQk.

Duhigg, C. (2012). How companies learn your secrets. *New York Times Magazine*. https://www.nytimes.com/2012/02/19/magazine/shopping-habits.html?pagewanted=all&_r=0. February 16.

Economist. (2017). Data is giving rise to a new economy. May 6.

Efron, B. and Hastie, T. (2016). *Computer Age Statistical Inference: Algorithms, Evidence and Data Science*. Cambridge University Press.

Einstein, A. n.d. Quote from http://www.azquotes.com/quote/811850.

Fienberg, S. (1979). Graphical methods in statistics. *American Statistician* 33(4): 165–178.

Fisher, R.A. (1922). On the mathematical foundations of theoretical statistics. *Philosophical Transactions of the Royal Society, Series A* 222: 309–368.

Godfrey, B. and Kenett, R.S. (2007). Joseph M. Juran, a perspective on past contributions and future impact. *Quality and Reliability Engineering International* 23: 653–663.

Greene, R. and Elffers, J. (1998). *The 48 Laws of Power*. Viking Penguin.

Hahn, G. (2003). The embedded statistician. Youden address. http://rube.asq.org/statistics/design-of-experiments/the-embedded-statistician.pdf.

Hahn, G.J. (2007). The business and industrial statistician: past, present and future. *Quality and Reliability Engineering International* 23: 643–650.

Hahn, G.J. and Doganaksoy, N. (2011). *A Career in Statistics: Beyond the Numbers*. Wiley.

Hartman, E., Grieve, R., Ramsahai, R. et al. (2015). From SATE to PATT: combining experimental with observational studies to estimate population treatment effects. *Journal of the Royal Statistical Society, Series A* 178: 757–778.

Harvard Business Review. (2013). Data and organizational issues reduce confidence. http://go.qlik.com/rs/qliktech/images/HBR_Report_Data_Confidence.PDF?sourceID1=mkto-2014-H1.

Harvard Business Review (2018). *HBR Guide to Data Analytics Basics for Managers*. Harvard Business Review Publishing.

Hastie, T., Tibshirani, R., and Friedman, J. (2009). *The Elements of Statistical Learning: Data Mining, Inference, and Prediction*, 2e. Springer.

Henke, N., Bughnn, J., Chui, M. et al. (2016). *The Age of Analytics: Competing in a Data Driven World*. McKinsey.

Huber, P. (2011). *Data Analysis: What Can Be Learned from the Past 50 Years*. Wiley.

Hunter, W.G. (1979). Private communication to first author, Madison, WI.

IDC. (2018). www.idc.com.

James, G., Witten, D., Hastie, T. et al. (2013). *An Introduction to Statistical Learning: With Applications in R*. Springer.

Joiner, B.L. (1985). The key role of statisticians in the transformation of North American industry. *American Statistician* 39(3): 233–234.

Joiner, B.L. (1994). *Fourth Generation Management: The New Business Consciousness*. New York: McGraw Hill.

Joint Commission International (2018). JCI accreditation standards for hospitals, 6e. https://www.jointcommissioninternational.org/jci-accreditation-standards-for-hospitals-6th-edition.

Juran, J.M. (1988). *Juran on Planning for Quality*. New York: Free Press.

Kaggle. (2017). The state of data science and machine learning. https://www.kaggle.com/surveys/2017.

Kahneman, D., Lovallo, D., and Sibony, O. (2011). The big idea: before you make that big decision. *Harvard Business Review*, February, pp. 50–60.

Kaplan, R. and Norton, D. (1996). Using the balanced scorecard as a strategic management system. *Harvard Business Review*, January–February, pp. 75–85.

Katkar, R. and Reiley, D.H. (2006). Public versus secret reserve prices in eBay auctions: results from a Pokémon field experiment. *Advances in Economic Analysis and Policy* 6(2): Article 7.

Kenett, R.S. (2008). From data to information to knowledge. *Six Sigma Forum Magazine*, November, pp. 32–33.

Kenett, R.S. (2015). Statistics: a life cycle view (with discussion). *Quality Engineering* 27(1): 111–129.

Kenett, R.S. (2017). The Information Quality Framework for Evaluating Manufacturing 4.0 Analytics Distinguished Lecture. TUE Data Science Center Eindhoven. https://assets.tue.nl/fileadmin/content/faculteiten/win/DSCE/Research/5Lecture_series/20170601_DSCe_Lecture_Ron_Kenett.pdf. June 1.

Kenett, R.S. and Baker, E. (2010). *Process Improvement and CMMI for Systems and Software: Planning, Implementation, and Management*. Taylor and Francis, Auerbach Publications.

Kenett, R.S., Coleman, S.Y., and Stewardson, D. (2003). Statistical efficiency: the practical perspective. *Quality and Reliability Engineering International* 19: 265–272.

Kenett, R.S., de Frenne, A., and Tort-Martotell, X. (2008). The statistical efficiency conjecture. In: *Statistical Practice in Business and Industry* (ed. S. Coleman, T. Greenfield, D. Stewardson, et al.), 61–95. Wiley.

Kenett, R.S., Pfeffermann, D., and Steinberg, D.M. (2018). Election polls – a survey, critique and proposals. *Annual Review of Statistics and Its Application* 5: 1–24. http://www.annualreviews.org/doi/abs/10.1146/annurev-statistics-031017-100204.

Kenett, R.S. and Raanan, Y. (2011). *Operational Risk Management: A Practical Approach to Intelligent Data Analysis*. Chichester, UK: Wiley.

Kenett, R.S. and Salini, S. (2011). *Modern Analysis of Customer Surveys with Applications Using R*. Wiley.

Kenett, R.S. and Shmueli, G. (2014). On information quality (with discussion).

Kenett, R.S. and Shmueli, G. (2016a). *Information Quality: The Potential of Data and Analytics to Generate Knowledge*. Wiley.

Kenett, R.S. and Shmueli, G. (2016b). Helping authors and reviewers ask the right questions: the InfoQ framework for reviewing applied research. *Statistical Journal of the International Association for Official Statistics (IAOS)* 32: 11–19.

Kenett, R.S. and Thyregod, P. (2006). Aspects of statistical consulting not taught by academia. *Statistica Neerlandica* 60(3): 396–412.

Kenett, R.S., Zacks, S., and contributions by Amberti, D. (2014). *Modern Industrial Statistics: With Applications in R, MINITAB and JMP*, 2e. Wiley.

Kohavi, R. and Thomke, S. (2017). The surprising power of online experiments. *Harvard Business Review*. https://hbr.org/2017/09/the-surprising-power-of-online-experiments. September–October.

Laney, D. (2017). *Infonomics: How to Monetize, Manage, and Measure Information as an Asset for Competitive Advantage*. Routledge.

Lavy, V. (2010). Effects of free choice among public schools. *Review of Economic Studies* 77(3): 1164–1191. https://academic.oup.com/restud/article-abstract/77/3/1164/1570661.

Lewis, M. (2017). *The Undoing Project: A Friendship That Changed Our Minds*. W.W. Norton.

Loftus, E.F. (2013). Eyewitness testimony in the Lockerbie bombing case. *Memory* 21: 584–590.

Lohr, S. (2018). Facial recognition is accurate, if you're a white guy. *New York Times*. https://www.nytimes.com/2018/02/09/technology/facial-recognition-race-artificial-intelligence.html. February 9.

Masic, I., Miokvic, M., and Muhamedagic, B. (2008). Evidence based medicine – new approaches and challenges. *Journal of Academy of Medical Sciences of Bosnia and Herzegovina* 16(4): 219–225. https://www.ncbi.nlm.nih.gov/pmc/articles/PMC3789163.

McAfee, A. and Brynjolffson, E. (2012). Big data: the management revolution. *Harvard Business Review*. https://hbr.org/2012/10/big-data-the-management-revolution. October.

McGarvie, M. and McElheran, K. (2018). Pitfalls of data-driven decisions. In: *HBR Guide to Data Analytics Basics for Managers*, 155–164. Harvard Business Review Press.

Nagle, T., Redman, T., and Sammon, D. (2017). Only 3% of companies' data meets basic quality standards. *Harvard Business Review*. https://hbr.org/2017/09/only-3-of-companies-data-meets-basic-quality-standards.

O'Keefe, K. and O'Brien, D. (2018). *Ethical Data and Information Management Concepts, Tools and Methods*. Kogan Page.

O'Neil, C. (2016). *Weapons of Math Destruction: How Big Data Increases Inequality and Threatens Democracy*. Crown.

Pearl, J. and Bareinboim, E. (2011). Transportability across studies: a formal approach. Technical Report R-372, Cognitive Systems Laboratory, Dept. Computer Science, Univerisity of California, Los Angeles.

Pearl, J. and Bareinboim, E. (2014). External validity: from do-calculus to transportability across populations. *Statistical Science* 29(4): 579–595.

Pirelli. (2016). How Pirelli is becoming data driven. https://www.pirelli.com/global/en-ww/life/how-pirelli-is-becoming-data-driven. March 23.

Pollack, A. (1999). Two teams, two measures equaled one lost spacecraft. *New York Times*. https://archive.nytimes.com/www.nytimes.com/library/national/science/100199sci-nasa-mars.html?scp=2. October 1.

Rao, C.R. (1985). Weighted distributions arising out of methods of ascertainment: what population does a sample represent? In: *A Celebration of Statistics: The ISI Centenary Volume* (ed. A.C. Atkinson and S.E. Fienberg), 543–569. New York: Springer.

Rasch, G. (1977). On specific objectivity: an attempt at formalizing the request for generality and validity of scientific statements. *Danish Yearbook of Philosophy* 14: 58–93.

Redman, T. (2008). *Data Driven: Profiting from Your Most Important Business Asset*. Harvard Business Review Press.

Redman, T. (2013a). What separates a good data scientist from a great one. *Harvard Business Review*. http://blogs.hbr.org/2013/01/the-great-data-scientist-in-fo. January 28.

Redman, T. (2013b). Are you data-driven? Take a hard look in the mirror. *Harvard Business Review*. http://blogs.hbr.org/2013/07/are-you-data-driven-take-a-har. July 11.

Redman, T. (2013c). Become more data-driven by breaking these bad habits. *Harvard Business Review*. http://blogs.hbr.org/2013/08/becoming-data-driven-breaking. August 12.

Redman, T. (2013d). Are you ready for a chief data officer? *Harvard Business Review*. https://hbr.org/2013/10/are-you-ready-for-a-chief-data-officer. October 30.

Redman, T. (2013e). How to start thinking like a data scientist. *Harvard Business Review*. https://hbr.org/2013/11/how-to-start-thinking-like-a-data-scientist. November 29.

Redman, T. (2014). Data doesn't speak for itself. *Harvard Business Review*. https://hbr.org/2014/04/data-doesnt-speak-for-itself. April 29.

Redman, T. (2015). Can your data be trusted? *Harvard Business Review*. https://hbr.org/2015/10/can-your-data-be-trusted. October 29.

Redman, T. (2016). *Getting in Front on Data: Who Does What*. Technics.

Redman, T. (2017a). The best data scientists get out and talk to people. *Harvard Business Review*. https://hbr.org/2017/01/the-best-data-scientists-get-out-and-talk-to-people. January 26.

Redman, T. (2017b). Root out bias from your decision-making process. *Harvard Business Review*. https://hbr.org/2017/03/root-out-bias-from-your-decision-making-process. March 17.

Redman, T. (2017c). Seizing opportunity in data quality. MIT Sloan Management Review. https://sloanreview.mit.edu/article/seizing-opportunity-in-data-quality. November 29.

Redman, T. (2018a). Are you setting your data scientists up to fail? *Harvard Business Review*. https://hbr.org/2018/01/are-you-setting-your-data-scientists-up-to-fail. January 25.

Redman, T. (2018b). If your data is bad, your machine learning tools are useless. *Harvard Business Review*. https://hbr.org/2018/04/if-your-data-is-bad-your-machine-learning-tools-are-useless. April 2.

Redman, T. and Sweeney, W. (2013a). To work with data, you need a lab and a factory. *Harvard Business Review*. https://hbr.org/2013/04/two-departments-for-data-succe. April 24.

Redman, T. and Sweeney, W. (2013b). Seven questions to ask your data geeks. *Harvard Business Review*. https://hbr.org/2013/06/seven-questions-to-ask-your-da. June 10.

Rosling, H. (2007). The best stats you've ever seen. https://www.youtube.com/watch?v=hVimVzgtD6w&t=43s.

Ross, C. (2017). IBM pitched its Watson supercomputer as a revolution in cancer care. It's nowhere close. https://www.statnews.com/2017/09/05/watson-ibm-cancer. September 5.

Rubin, D. (1987). *Multiple Imputation for Nonresponse in Surveys*. New York: Wiley.

Schmarzo, B. (2017). Unintended consequences of the wrong measures. https://www.datasciencecentral.com/profiles/blogs/unintended-consequences-of-the-wrong-measures?_ga=2.21159096.226186582.1522766955-1679466071.1522672939. October 19.

Senge, P. (1990). *The Fifth Discipline: The Art and Practice of the Learning Organization*. New York: Doubleday.

Shannon, C. (1948). A mathematical theory of communication. *Bell System Technical Journal* 27: 379–423 and 623–656.

Shewhart, W.A. (1926). Quality control charts. *Bell System Technical Journal* 5: 593–603.

Shmueli, G. (2010). To explain or to predict? *Statistical Science* 25(3): 289–310.

Shmueli, G. (2017). Analyzing behavioral big data: methodological, practical, ethical, and moral issues. *Quality Engineering* 29(1): 57–74.

Shmueli, G. (2018). Data ethics regulation: two key updates in 2018. http://www.bzst.com/2018/02/data-ethics-regulation-two-key-updates.html.

Silver, N. (2012). *The Signal and the Noise: Why So Many Predictions Fail*. Penguin Press.

Simon, D. (2010). Selective attention test. https://www.youtube.com/watch?v=vJG698U2Mvo.

Singer, M. (2018). Tech's ethical "dark side": Harvard, Stanford and others want to address it. *New York Times*. https://www.nytimes.com/2018/02/12/business/computer-science-ethics-courses.html. February 12.

Srivastiva, D., Caannapieco, M., and Redman, T. (2019). Ensuring high-quality private data for responsible data science: vision and challenges. *Journal of Data and Information Quality*. Accepted for publication.

Surveytown. (2016). 10 examples of biased survey questions. https://surveytown.com/10-examples-of-biased-survey-questions. April 12.

Taguchi, G. (1987). *Systems of Experimental Design*, vol. 1–2 (ed. D. Clausing). New York: UNIPUB/Kraus International Publications.

Tashea, J. (2017). Courts are using AI to sentence criminals. That must stop now. *Wired*. https://www.wired.com/2017/04/courts-using-ai-sentence-criminals-must-stop-now. April 17.

Tufte, E.R. (1997). *Visual Explanations: Images and Quantities, Evidence and Narrative*. Cheshire, CT: Graphics Press.

Tukey, J.W. (1962). The future of data analysis. *Annals of Mathematical Statistics* 33: 1–67.

Tversky, A. and Kahneman, D. (1981). The framing of decisions and the psychology of choice. *Science* 211(4481): 453–458.

van Buuren, S. (2012). *Flexible Imputation of Missing Data*. Chapman & Hall/CRC.

Wang, S., Jank, W., and Shmueli, G. (2008). Explaining and forecasting online auction prices and their dynamics using functional data analysis. *Journal of Business Economics and Statistics* 26: 144–160.

Wikipedia. (2018a). Imputation (statistics). https://en.wikipedia.org/wiki/Imputation_(statistics).

Wikipedia. (2018b). Response bias. https://en.wikipedia.org/wiki/Response_bias.

Wikipedia. (2018c). Truthiness. https://en.wikipedia.org/wiki/Truthiness.

Wilder-James, E. (2016). Breaking down data silos. *Harvard Business Review*. https://hbr.org/2016/12/breaking-down-data-silos. December 5.

Zhang, M.X., Van Den Brakel, J., Honchar, O. et al. (2018). Using state space models for measuring statistical impacts of survey redesigns. Australian Bureau of Statistics Research Paper. 1351.0.55.160. www.abs.gov.au/ausstats/abs@.nsf/mf/1351.0.55.160?OpenDocument.

A List of Useful Links

Videos, Blogs, and Presentations

1. Information quality in the golden age of analytics. JMP analytically speaking series. Interview (https://www.jmp.com/en_us/events/ondemand/analytically-speaking/quality-assurance-in-the-golden-age-of-analytics.html) and blog post (https://community.jmp.com/t5/JMP-blog-authoring/Information-quality-in-the-golden-age-of-analytics/ba-p/66903).
2. From quality by design (QbD) to information quality (InfoQ): a journey through science and business analytics. Plenary talk at JMP Summit Prague. Starts at 5:20 (https://community.jmp.com/t5/Discovery-Summit-Europe-2017/Plenary-Session-From-Quality-by-Design-to-Information-Quality-A/ta-p/37537).
3. The Real Work of Data Science: How to Turn Data into Information, Better Decisions, and Stronger Organizations, The ENBIS2018 Box Medal Lecture. https://videos.univ-lorraine.fr/index.php?act=view&id=6456&fbclid=IwAR3ZnR-
4. The SPCLive system for statistical process control in assembly production. https://www.youtube.com/watch?v=E7W99sCYYos. See also https://vimeo.com/285180512.
5. A life cycle view of statistics. JSM18 panel. https://blogisbis.wordpress.com/2018/08/21/panel-discussion-on-a-life-cycle-view-of-statistics-at-the-jsm-2018.
6. What does big data mean for business? https://www.youtube.com/watch?v=qEm1ngJDOlE&t=221s.
7. Two moments that matter in data. https://www.youtube.com/watch?v=dBpHHAG_d9E.
8. If your data is bad, your machine learning tools are useless. Video #1 (https://www.youtube.com/watch?v=efRKika4SOE&t=10s), video #2 (https://www.youtube.com/watch?v=woaL40K6xyA), video #3 (https://www.youtube.com/watch?v=rnjmlc4MekI&t=23s), video #4 (https://www.youtube.com/watch?v=HS8SHzo9Vvk), video #5 (https://www.youtube.com/watch?v=AGSpsZDgzRE&t=9s), video #6 (https://www.youtube.com/watch?v=CdlM1QQGVXg&t=83s), video #7 (https://www.youtube.com/watch?v=iXw6f_mfyCo&t=1s).

The Real Work of Data Science: Turning Data into Information, Better Decisions, and Stronger Organizations, First Edition. Ron S. Kenett and Thomas C. Redman.
© 2019 Ron S. Kenett and Thomas C. Redman. Published 2019 by John Wiley & Sons Ltd.
Companion website: www.wiley.com/go/kenett-redman/datascience

9. The Friday afternoon measurement for data quality. https://www.youtube.com/watch?v=X8iacfMX1nw&t=22s.
10. Will bad data make bad robots? https://www.youtube.com/watch?v=JgReisW0hBw.
11. An outlook on data science. https://mathesia.com/home/Mathesia_Outlook_2019.pdf.

Articles

Data Science in Global Companies

"Data science at Alibaba."
https://blogisbis.wordpress.com/2017/11/16/data-science-at-alibaba.
"How Pirelli is becoming data driven."
https://www.pirelli.com/global/en-ww/life/how-pirelli-is-becoming-data-driven.
"Why you're not getting value from your data science." December 7, 2016.
https://hbr.org/2016/12/why-youre-not-getting-value-from-your-data-science.

On Deep Learning and Artificial Intelligence
"Artificial intelligence pioneer says we need to start over."
https://www.axios.com/artificial-intelligence-pioneer-says-we-need-to-start-over-1513305524-f619efbd-9db0-4947-a9b2-7a4c310a28fe.html.
"Deep learning: a critical appraisal."
https://arxiv.org/abs/1801.00631.
"Getting value from machine learning isn't about fancier algorithms – it's about making it easier to use." March 6, 2018.
https://hbr.org/2018/03/getting-value-from-machine-learning-isnt-about-fancier-algorithms-its-about-making-it-easier-to-use.
"IBM pitched its Watson supercomputer as a revolution in cancer care. It's nowhere close." September 5, 2017.
https://www.statnews.com/2017/09/05/watson-ibm-cancer.

Data and Data Integration
"A Cambridge professor on how to stop being so easily manipulated by misleading statistics."
https://qz.com/643234/cambridge-professor-on-how-to-stop-being-so-easily-manipulated-by-misleading-statistics.
"Data is not information."
https://www.technologyreview.com/s/514591/the-dictatorship-of-data/?_ga=2.105454147.13
11780712.1522672939-1679466071.1522672939.
"Divide & recombine (D&R) with DeltaRho for big data analysis."
https://ww2.amstat.org/meetings/sdss/2018/onlineprogram/AbstractDetails.cfm?AbstractID=
304537&utm_source=informz&utm_medium=email&utm_campaign=asa&_
zs=IVpOe1&_zl=TX2b.4.
"Social media big data integration: a new approach based on calibration."
https://www.sciencedirect.com/science/article/pii/S0957417417308667.

Advanced Manufacturing, Vegetation, and Global Warming
"A road map for applied data sciences supporting sustainability in advanced manufacturing: the information quality dimensions."
https://www.sciencedirect.com/science/article/pii/S2351978918301392.

"Vegetation intensity throughout the year for Africa."
https://www.reddit.com/r/dataisbeautiful/comments/8oo1ah/vegetation_intensity_throughout_
the_year_for.
"What's really warming the world." June 24, 2015.
https://www.bloomberg.com/graphics/2015-whats-warming-the-world.

On Statistics and Academia
"Academics can change the world – if they stop talking only to their peers."
https://theconversation.com/academics-can-change-the-world-if-they-stop-talking-only-to-their-
peers-55713?utm_source=twitter&utm_medium=twitterbutton.
**"For survival, statistics as a profession needs to provide added value to fellow scientists
or customers in business and industry."**
http://www.statisticsviews.com/details/feature/4812131/For-survival-statistics-as-a-
profession-needs-to-provide-added-value-to-fellow-s.html.
**"Psychology journal editor asked to resign for refusing to review papers unless he can
see the data."**
https://boingboing.net/2017/03/02/psychology-journal-editor-aske.html.

On p Values
"p hacking."
https://www.methodspace.com/primer-p-hacking/?_ga=2.153362040.1311780712.15226729
39-1679466071.1522672939.
"'To p or not to p' – my thoughts on the ASA Symposium on Statistical Inference."
https://blogisbis.wordpress.com/2017/10/24/to-p-or-not-to-p-my-thoughts-on-the-asa-
symposium-on-statistical-inference.

Experiments
"'I placed too much faith in underpowered studies:' Nobel Prize winner admits mistakes."
https://retractionwatch.com/2017/02/20/placed-much-faith-underpowered-studies-nobel-
prize-winner-admits-mistakes.
"In a big data world, don't forget experimentation." May 8, 2013.
http://blogs.hbr.org/2013/05/in-a-big-data-world-dont-forge.
"A refresher on randomized controlled experiments." March 30, 2017.
https://hbr.org/2016/03/a-refresher-on-randomized-controlled-experiments.
"The surprising power of online experiments." September 2017.
https://hbr.org/2017/09/the-surprising-power-of-online-experiments.

Statistical Potpourri
"Censor bias."
https://medium.com/@penguinpress/an-excerpt-from-how-not-to-be-wrong-by-
jordan-ellenberg-664e708cfc3d?_ga=2.119218121.1311780712.1522672939-1679466071.
1522672939.
"Cherry picking."
https://www.economicshelp.org/blog/21618/economics/cherry-picking-of-data/?_ga=2.1
10836549.1311780712.1522672939-1679466071.1522672939.

"The Hawthorne effect."
https://www.economist.com/node/12510632?_ga=2.182730470.1311780712.1522672939-1
   679466071.1522672939.

"Overfitting."
https://www.kdnuggets.com/2014/06/cardinal-sin-data-mining-data-science.html?_ga=2.
   144900084.1311780712.1522672939-1679466071.1522672939.

"A refresher on regression analysis." November 4, 2015.
https://hbr.org/2015/11/a-refresher-on-regression-analysis.

"A refresher on statistical significance." February 16, 2016.
https://hbr.org/2016/02/a-refresher-on-statistical-significance.

"Selection bias."
https://www.khanacademy.org/math/statistics-probability/designing-studies/sampling-and-
   surveys/a/identifying-bias-in-samples-and-surveys?_ga=2.178468836.1311780712.15226
   72939-1679466071.1522672939.

"Simpson's paradox."
https://www.brookings.edu/blog/social-mobility-memos/2015/07/29/when-average-isnt-good-
   enough-simpsons-paradox-in-education-and-earnings/?_ga=2.114356423.1311780712.
   1522672939-1679466071.1522672939.

"Spurious correlations."
http://www.tylervigen.com/spurious-correlations?_ga=2.80296535.1311780712.1522672939-
   1679466071.1522672939.

Index

The Real Work of Data Science: Turning Data into Information, Better Decisions, and Stronger Organizations,
First Edition. Ron S. Kenett and Thomas C. Redman.
© 2019 Ron S. Kenett and Thomas C. Redman. Published 2019 by John Wiley & Sons Ltd.
Companion website: www.wiley.com/go/kenett-redman/datascience